MCDONNELL DOUGLAS
F-4 PHANTOM II
AIR SUPERIORITY LEGEND

A USS *John F. Kennedy*-based VF-32 F-4B Phantom II performs fleet-defence duty on 16 January 1971. (US National Archives at College Park, Maryland, Still Pictures Branch)

MCDONNELL DOUGLAS F-4 PHANTOM II
AIR SUPERIORITY LEGEND

MARK A. CHAMBERS

The History Press

A US Navy F-4B Phantom II intercepts a Soviet Tupolev Tu-95 Bear bomber over the Pacific Ocean in May 1971. (US National Archives at College Park, Maryland, Still Pictures Branch)

First published 2018

The History Press
The Mill, Brimscombe Port
Stroud, Gloucestershire, GL5 2QG
www.thehistorypress.co.uk

British Library Cataloguing in Publication Data.
A catalogue record for this book is available from the British Library.

ISBN 978 0 7509 8279 5

Typesetting and origination by The History Press
Printed in India by Thomson Press

Cover illustrations
Front: A USAF 3rd TFS F-4E in flight during exercise Cope North '81–4.
(US National Archives at College Park, Maryland, Still Pictures Branch);
Back: A Royal Navy 892 Naval Air Squadron Phantom II FG.1 preparing
for flight operations aboard HMS *Ark Royal* on 2 March 1972. (US Navy)

CONTENTS

ACKNOWLEDGEMENTS

Numerous individuals deserve great thanks for providing crucial support for the completion of this book. First and foremost, great thanks go to my loving family – my wife Lesa, daughter Caitlyn, and sons Patrick and Ryan – for tolerating my ceaseless words of enthusiasm and providing encouragement and support for this project. Great thanks also go to David Pfeiffer (Civil Records Archivist), Nate Patch (Military Records Archivist) and the entire staff of the Textual Reference Branch of the US National Archives and Record Administration (NARA) at College Park, Maryland; and to Holly Reed and the entire staff of the Still Pictures Branch of the US NARA at College Park, Maryland. Finally, great thanks go to Amy Rigg, Commissioning Editor at The History Press, for her unwavering and fantastic encouragement and support in seeing this project through to publication.

FURTHER READING

Chambers, Joseph R. (2000) NASA SP-2000-4519, *Partners in Freedom: Contributions of the Langley Research Center to U.S. Military Aircraft of the 1990's*, Monographs in Aerospace History Number 19, The NASA History Series, National Aeronautics and Space Administration, Washington, DC.

Davies, Peter E. *USAF F-4 Phantom II MiG Killers 1972–73* (Osprey Combat Aircraft 55). Botley, Oxford, UK; New York, NY, USA: Osprey Publishing Limited, 2005.

Elward, Brad and Peter Davies. *US Navy F-4 Phantom II MiG Killers 1965–70* (Osprey Combat Aircraft 26). Botley, Oxford, UK: Osprey Publishing Limited, 2001.

Grossnick, Roy and William J. Armstrong. *United States Naval Aviation, 1910–1995*. Annapolis, Maryland: Naval Historical Center, 1997.

McCarthy, Donald J., Jr. *USAF F-4 and F-105 MiG Killers of the Vietnam War 1965–1973*. Atglen, PA: Schiffer Military History, 2005.

McCarthy, Donald J., Jr. MiG Killers: *A Chronology of U.S. Air Victories in Vietnam 1965–1973*. North Branch, MN: Specialty Press, 2009.

Richardson, Doug and Mike Spick. (1984) *Modern Fighting Aircraft, Volume 4: F-4*. Published in United States by Arco Publishing Inc., New York, NY. Published in United Kingdom by Salamander Books Ltd., London.

Thornborough, Anthony M. and Peter E. Davies. *The Phantom Story*. London: Arms and Armour Press, 1994.

INTRODUCTION

The McDonnell Douglas F-4 Phantom II has endured as a legendary icon of American airpower in the early- to mid-1960s, the Vietnam War, the Cold War and Operation Desert Storm. The aircraft proved to be a tough, rugged and excellent all-weather, air superiority fighter/bomber that performed admirably in the Vietnam War and helped secure Allied Coalition victory in Operation Desert Storm. From flying fleet defence missions during the Cuban Missile Crisis and performing as an excellent air superiority fighter/bomber during the Vietnam War, to taking out Iraqi surface-to-air missile (SAM) sites in the 'Wild Weasel' role during Operation Desert Storm, the Phantom II did it all. Other nations have also benefited tremendously from the acquisition of export versions of this remarkable aircraft. This book tells the fascinating story of McDonnell's aircraft design philosophy as well as this truly unique aircraft's design and development. It also focuses on the Phantom II's glorious record in the Vietnam War, various Arab-Israeli conflicts, the Cold War and Operation Desert Storm.

1

MCDONNELL AIRCRAFT CORPORATION/MCDONNELL DOUGLAS CORPORATION

The McDonnell Aircraft Corporation rose from meagre beginnings to major prominence in the United States aerospace industry in a relatively short period of time during the twentieth century. The highly successful aerospace company later merged with Douglas to form an American aviation giant, McDonnell Douglas. McDonnell Aircraft was established by James Smith McDonnell on 6 July 1939 in St Louis, Missouri. McDonnell had previously ventured, unsuccessfully, into the American aviation industry with his own company, J.S. McDonnell & Associates, in Milwaukee, Wisconsin, in 1928. McDonnell's primary pursuit was the production of a general aviation aircraft that could be flown by a family. His dream, however, was crushed by the Great Depression, which forced him out of business. After being employed by Glenn L. Martin, McDonnell resigned in 1938 to revisit and pursue his corporate aviation

dream. He subsequently established his own company, McDonnell Aircraft Corporation, in St Louis, Missouri, in 1939.

The McDonnell Aircraft Corporation began to hit its stride during the Second World War. When the company commenced operations in 1939, it employed a staff of just fifteen employees; by the war's end, it employed 5,000 people. In addition to serving as a prominent aircraft-part production company, McDonnell's primary focus during the Second World War was on the development and production of the experimental XP-67 Bat fighter. Among McDonnell's additional pursuits was the LBD-1 Gargoyle guided missile. Following the Second World War, the company languished as a result of the termination of government contracts and an overabundance of aircraft. Consequently, the company was forced to lay off a significant portion of its staff. Nevertheless, the

The experimental McDonnell XP-67 Bat fighter, designed for the USAAF. (US National Archives at College Park, Maryland, Textual Reference Branch)

An in-flight view of the experimental McDonnell XP-67 Bat fighter in 1944. (US National Archives at College Park, Maryland, Textual Reference Branch)

The McDonnell XFD-1 (prototype of the famous FD-1 Phantom series) at the Naval Air Test Center Patuxent River, Maryland, on 8 May 1946. The XFD-1 was McDonnell's first Phantom jet fighter endeavour. (US National Archives at College Park, Maryland, Textual Reference Branch)

A side view of the McDonnell XFD-1 Phantom at the Naval Air Test Center Patuxent River, Maryland, on 8 May 1946. (US National Archives at College Park, Maryland, Textual Reference Branch)

A McDonnell F2H Banshee being serviced at Naval Air Test Center Patuxent River, Maryland, during the late 1940s. (US National Archives at College Park, Maryland, Still Pictures Branch)

The McDonnell XF3H Demon prototype at Naval Air Test Center Patuxent River, Maryland, on 11 August 1953. (US National Archives at College Park, Maryland, Still Pictures Branch)

A McDonnell F3H-1N Demon at Naval Air Station Patuxent River, Maryland, on 17 September 1954. (US National Archives at College Park, Maryland, Still Pictures Branch)

A NACA McDonnell F-101A Voodoo interceptor at NACA Langley Memorial Aeronautical Laboratory in Hampton, Virginia, in 1956. (US National Archives at College Park, Maryland, Still Pictures Branch)

A USAF McDonnell RF-101A Voodoo on a photoreconnaissance sortie above Vietnam in May 1967. (USAF)

A 123rd Fighter-Interceptor Squadron Redhawks, 142nd Fighter Interceptor Group, Oregon Air National Guard McDonnell F-101B Voodoo interceptor in flight during the 1970s. (USAF)

dawning of the Jet Age and the beginning of the Korean War propelled McDonnell's military fighter-aircraft production efforts.

McDonnell's jet fighter pursuits began in 1946 with its development of the FD-1 Phantom, which became an instant success. Follow-up successes were experienced with the company's development of the F2H Banshee (used during the Korean War), F3H Demon and F-101 Voodoo. The Century Series F-101 Voodoo saw service with the United States Air Force (USAF) and combat duty in an aerial photoreconnaissance capacity during the Cuban Missile Crisis in October 1962 and the Vietnam War.

When the Great Space Race began following the launch and success of the Soviet Union's Sputnik in 1957, McDonnell answered the National Aeronautics and Space Administration's (NASA's) demands for advanced space capsules for projects Mercury and Gemini. Advancements made in both the hardware and software systems (specifically, a space-flight computer), as well as the rendezvous and docking capability, for the Gemini spacecraft enabled the United States to ultimately overtake the Soviet Union in the Great Space Race. These technological breakthroughs would serve as stepping stones to the United States' ultimate triumph, the Apollo Program.

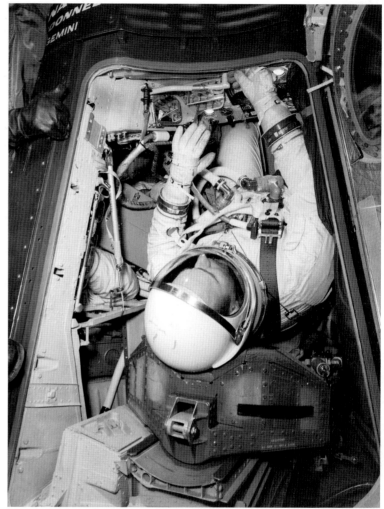

Gemini VI astronaut Thomas P. Stafford performs pre-flight checks on board the Gemini VI space capsule before lift-off on 15 December 1965. McDonnell built the Gemini space capsules. (US National Archives at College Park, Maryland, Still Pictures Branch)

Astronaut John Glenn enters the *Friendship 7* Mercury space capsule on 20 February 1962. McDonnell built the Mercury space capsules. (NASA)

On 15 December 1965, Gemini VI (from which this photo was taken) successfully rendezvoused with Gemini VII in Earth orbit. McDonnell built both space capsules. (NASA)

McDonnell Douglas F-15 Eagles. (US National Archives at College Park, Maryland, Still Pictures Branch)

McDonnell Douglas AV-8B Harrier II. (US National Archives at College Park, Maryland, Still Pictures Branch)

However, McDonnell still suffered during times of peace due to a lack of a civilian aircraft production division; therefore, in 1963, it began seeking a partner with a strong civil aviation production division and past. McDonnell found such a partner in Douglas Aircraft. On 28 April 1967, the two US aerospace corporate giants merged to form the McDonnell Douglas Corporation (MDC), with its headquarters and main base of operations located in St Louis, Missouri. Notable military aircraft produced by MDC included the F-15 Eagle, AV-8B Harrier II, F/A-18 Hornet and F/A-18E/F Super Hornet. In August 1997, McDonnell Douglas merged with Boeing. Today, the Boeing Defense and Space Division's main base of operations is in St Louis, Missouri.

McDonnell Douglas/Boeing F/A-18E Super Hornet. (US Navy photo by Photographer's Mate 2nd Class Angela M. Virnig)

McDonnell Douglas/Boeing F/A-18F Super Hornet. (US Navy photo by Greg L. Davis)

2

PHANTOM II DESIGN AND DEVELOPMENT

In 1946, Dave Lewis was hired as chief of aerodynamics at McDonnell. Lewis spearheaded the design and development of an advanced air superiority fighter/bomber that became known as the F-4 Phantom II in 1954. He rose to the rank of executive vice president of the company in 1958, ultimately rising to the rank of president and chief operating officer of McDonnell in 1962 and serving as president of McDonnell Douglas in 1969.

During the early 1950s, McDonnell identified the fact that the US Navy (USN) was seeking an advanced fighter/bomber to perform the roles of both an air superiority fighter and an attack bomber. In 1953, McDonnell simply decided to upgrade its F3H Demon fighter. Consequently, the company produced several versions of the Demon featuring engine upgrades – one version that utilised a Wright J67

engine and other versions that utilised either two Wright J65 engines or two General Electric J79 engines. Early tests indicated that the advanced Demon featuring the J79 powerplants was capable of achieving a maximum speed of Mach 1.97. McDonnell proposed its advanced Demon concept to the USN on 19 September 1953. The aircraft was designed to be interchangeable, featuring either one- or two-seat forward fuselages. Although the navy desired the new F3H-G/H mockup, it decided to stick with the new Grumman XF9F-9 Cougar and Vought XF8U-1 Crusader as its supersonic fighters.

As a result, McDonnell worked feverishly to redesign its advanced Demon, making it an all-weather fighter-bomber. On 18 October 1954, McDonnell received the 'green light' from the navy for the production of two YAH-1 prototypes. In mid 1955, the navy once again changed the requirements for the dogged fighter-bomber, now requesting that the aircraft serve as an all-weather fleet defender, featuring a 'back-seater' or Radar Intercept Officer (RIO).

Redesign and development work resulted in the XF4H-1 prototype. The aircraft was to be armed with four semi-recessed AAM-N-6 Sparrow III radar-guided air-to-air missiles. For maximum propulsion performance, the aircraft was fitted with two J79-GE-8 engines, complete with modulated afterburners. Using these engines, the aircraft could achieve a top speed of Mach 2.4. Due to lateral instability discovered in wind-tunnel tests, the aircraft's wings were readjusted with a 5° dihedral. The aircraft also featured specially designed wings for enhanced control at high angles of attack. In addition to an all-moving tailplane, the aircraft was equipped with the powerful AN/APQ-50 radar for all-weather fleet defence duty.

The McDonnell F3H-G/H advanced Demon mock-up. (US Navy)

The McDonnell F4H-1 at McDonnell Aircraft Corporation, St. Louis, Missouri, on 5 June 1958. The F4H-1 was capable of carrying both conventional and nuclear weapons. (Naval History and Heritage Command)

Father figures in the F-4 developmental programme, from left to right: David Lewis, Robert Little and Herman Barkey. (USAF)

The F4H-1 undergoes pre-flight checks by McDonnell ground crews at Edwards AFB, California, on 13 September 1959. (US National Archives at College Park, Maryland, Still Pictures Branch)

The navy requested the production of two XF4H-1 prototypes and five YF4H-1 pre-production aircraft on 25 July 1955. Piloted by Robert C. Little, the XF4H-1 successfully performed its first flight on 27 May 1958, although technical difficulties prevented landing gear retraction. The problem was solved, but additional flight testing led to several more aerodynamic refinements to the aircraft's overall design, including air intake enhancement and the addition of splitter plates in production variants that redirected the boundary layer away from the engine intakes. The XF4H-1 competed against the XF8U-3 Crusader III in fly-offs in 1958 and ultimately won the competition in December of the same year. The F4H successfully completed carrier feasibility flight trials, conducted aboard the USS *Independence*, on 15 February 1960. McDonnell and the military ultimately settled on the name 'Phantom II' for the new aircraft.

The Phantom II was capable of attaining a maximum speed in excess of Mach 2.2, and, on its way to achieving glory in combat, set numerous world speed and altitude records. After all was said and done, the Phantom II amassed an impressive tally of sixteen world records. With the exception of Skyburner, production variant Phantom IIs set these records. A total of five F-4 speed records stood for several years until they

Commander D.D. Engen boards the F4H-1 prior to heading out on a test flight on 13 September 1959. (US National Archives at College Park, Maryland, Still Pictures Branch)

The F4H-1 performs a take off on 13 September 1959. (US National Archives at College Park, Maryland, Still Pictures Branch)

were broken by the McDonnell Douglas F-15 Eagle in the mid 1970s. The first speed/altitude record was set during Operation Top Flight on 6 December 1959, when Commander Lawrence E. Flint, Jr., piloted the second XF4H-1 prototype to an altitude of 90,000ft (27,430m), while attaining a speed of Mach 2.5 in the skies above Edwards AFB, California. On 5 September 1960, an F-4H, piloted by USMC Lieutenant Colonel Thomas H. Miller, achieved a top speed of 1,216.76mph while flying a 500km closed circuit above Edwards AFB. The feat, in which the aircraft reached a speed of Mach 2.10, shattered the old speed record. On 25 September of the same year, an F-4H, piloted by Commander J.F. Davis, set a new speed record of Mach 2.31 while flying a 100km closed circuit at Edwards AFB. In commemoration of the fiftieth anniversary of naval aviation, Operation LANA (the L represented the Roman numeral 50, while ANA represented Anniversary of Naval Aviation) was conducted on 24 May 1961, in which Phantom IIs performed transcontinental flights across the United States in record-setting times. The Phantom IIs all performed these flights in less than 3 hours, but required some aerial refuelling from tankers. The fastest of these flights was made in 2 hours 47 minutes by an aircraft piloted by USN Lieutenant Richard Gordon, who later became a NASA Astronaut, and RIO USN Lieutenant Bobbie Young. Both airmen were later awarded the 1961 Bendix Trophy for their aerial feat; their cruising speed during the record-setting flight was in excess of 873mph. During Operation Sageburner on 28 August 1961, a F4H-1F set a world record for flying at an average speed of 902.714mph along a 3-mile course and at an altitude under 125ft. This feat was accomplished by pilot Lieutenant Huntington Hardisty and RIO Lieutenant Earl DeEsch. During Operation Skyburner on 22 November 1961, a special water-injection-equipped Phantom II, flown by Lieutenant Colonel Robert B. Robinson, established yet another Phantom II world record when the aircraft achieved an average speed of 1,606.342mph along a 20-mile (32.2km) two-way straight course. Also during the same operation, a different Phantom II established an altitude record of 66,443.8ft (20,252m) on 5 December 1961. At the beginning of 1962, several time-to-altitude records were established by a Phantom II during Operation High Jump. These records included 34.523 seconds to 9,840ft (3,000m) by Lieutenant Commander John Young, 48.787 seconds to 19,700ft (6,000m) and 61.629 seconds to 29,500ft (9,000m) by Commander David Langton, 77.156 seconds to 39,400ft (12,000m) by Lieutenant Colonel W. McGraw, 114.548 seconds to 49,200ft (15,000m) by Lieutenant Commander Del Nordberg,

On 6 December 1959, Commander Lawrence E. Flint, Jr. piloted the second XF4H-1 prototype to an altitude of 90,000ft and speed of Mach 2.5 at Edwards AFB during Operation Top Flight. The feat was the first speed/altitude record set by the Phantom II. (US National Archives at College Park, Maryland, Still Pictures Branch)

178.5 seconds to 65,600ft (20,000m) by Lieutenant Commander Taylor Brown, 230.44 seconds to 82,000ft (25,000m) by Lieutenant Commander John Young, and 371.43 seconds to 98,400ft (30,000m) by Lieutenant Commander Del Nordberg.[1]

Initial production Phantom II variants received Westinghouse AN/APQ-72 and AN-APG-50 radar upgrades, as well as canopy and cockpit enhancements. Following Defense Secretary Robert McNamara's request that the military adopt one fighter type for all branches, the USAF acquired its first Phantom IIs. In flight trials with a Convair F-106 Delta Dart, dubbed 'Operation Highspeed', a USN F-4B bested the Convair product, convincing the USAF to borrow two Navy F-4Bs. These aircraft were designated F-110As in January 1962. Following McNamara's F-4 variant unification decree in September 1962, all Navy Phantom IIs were designated F-4Bs and all USAF Phantom IIs F-4Cs. The initial USAF F-4C, equipped with more adaptable J79-GE-15 engines, successfully performed its first flight on 27 May 1963, easily achieving a speed in excess of Mach 2.

An F4H on the return taxi, with drag chute deployed, on 23 March 1960. (US National Archives at College Park, Maryland, Still Pictures Branch)

An F4H-1F Phantom II performing carrier trials aboard the USS *Independence* in April 1960. (US Navy)

USMC Lieutenant Colonel Thomas H. Miller prior to breaking the 500km speed record in an F4H-1. (US National Archives at College Park, Maryland, Still Pictures Branch)

On 21 October 1960, US President Dwight D. Eisenhower paid a visit to US Naval Air Station, North Island, San Diego, California, where he was shown an F4H-1, with the assistance of F4H-1 pilot Commander Tom Mie. (US National Archives at College Park, Maryland, Still Pictures Branch)

An F-4H performs flight and fleet acceptance trials above the Naval Air Test Center Patuxent River, Maryland, on 16 February 1961. (US National Archives at College Park, Maryland, Still Pictures Branch)

An F4H-1 in 1961. (US National Archives at College Park, Maryland, Still Pictures Branch)

An F4H-1 Phantom II performs an in-flight weapon test of a Sparrow radar-guided air-to-air missile on 5 June 1961. (US National Archives at College Park, Maryland, Still Pictures Branch)

An F4H-1 Phantom II, armed with iron bombs, on a weapons test flight in 1961. (US National Archives at College Park, Maryland, Still Pictures Branch)

A VF-101 Det. A Grim Reapers McDonnell F4H-1F Phantom II undergoes in-flight aerial refuelling from a US Navy Heavy Attack Squadron VAH-9 Hoot Owls A-3 Skywarrior during a Project LANA record-setting transcontinental flight in 1961. (US Navy)

In late January 1962, an F-110A (F-4C) departed McDonnell Aircraft Corporation, St. Louis, Missouri, for Tactical Air Command Headquarters, Langley AFB, Virginia. At Langley, the aircraft participated in orientation and evaluation studies. (US National Archives at College Park, Maryland, Still Pictures Branch)

An F4H-1 Phantom II performs another time-to-climb record-setting Project High Jump flight above Naval Air Station Point Mugu, California, on 31 March 1962. (US National Archives at College Park, Maryland, Still Pictures Branch)

Another view of the Langley AFB-bound F-110A (F-4C) departing McDonnell Aircraft Corporation, St Louis, Missouri, in late January 1962. (US National Archives at College Park, Maryland, Still Pictures Branch)

On 24 January 1962, two F-110As (F-4Cs) arrived at Tactical Air Command Headquarters, Langley AFB, Virginia, to participate in 120-day-long orientation and evaluation studies. (US National Archives at College Park, Maryland, Still Pictures Branch)

A rear view of an F-110A (F-4C) at Langley AFB, Virginia, on 24 January 1962. (US National Archives at College Park, Maryland, Still Pictures Branch)

At a Pentagon press conference, held on 5 February 1963, Lieutenant General Gabriel P. Disosway, USAF, Deputy Chief of Staff (DCS)/ Programs and Requirements, discussed the true essence of the McDonnell F-4 Phantom II:

Yesterday the Air Force took delivery of the first of 29 F4B Phantom II jet fighters on loan to us from the US Navy. These jets will be used during the spring and summer months to train instructor pilots and maintenance crews for the McDonnell F4C aircraft we will begin receiving this fall.

This procedure is rather unusual, so unusual I thought it worthy of your attention.

The Phantom II has probably had more official designations than any jet in the Armed Forces. It has been known as the F4H, F-110, F4B and F4C. From now on you can think of the 'B' model being Navy and the 'C' as Air Force.

The Phantom is a more easily used term as I will stick to that this morning as much as possible.

The Phantom is a Navy aircraft, born and bred. It was designed to be carrier operated and is the best in the world. The Air Force found it to be very adaptable to the tactical fighter mission.

The mission of a tactical fighter is very exacting as we see it in the Air Force. We require more from these jets and their pilots than any other weapon system in the Air Force. We expect the pilot to be gunner, bombardier, navigator, radio operator and perform a half dozen other tasks. We expect his airplane to be an air superiority fighter, (day or night), a dive bomber, a few altitude attack aircraft. We let it range over the entire tactical field – that of close air support, interdiction and air superiority.

There is a very good reason for all this. Based on our experience in several wars we realize it is impossible to purchase an airplane and train a pilot for each of these missions. Not only would that be prohibitive financially but it would be a waste of national resources for these missions are not always carried out simultaneously. If we used different aircraft for each, it is possible part of the Air Force would be standing down while the air superiority portion needed help to win the air battle and make it possible for the close air support team to fly over the battlefield.

What we needed was one airplane and pilot who could accomplish all three jobs. This is where we get the title of 'tactical fighter'. It comes from the melding of the functions of the day and night fighter and the fighter-bomber in one airplane.

A year ago we were searching for another tactical fighter to add to the USAF inventory to modernize Tactical Air Command as fast as possible. We needed a second production source to augment the F-105 Thunderchief, which, I would like to point out, was the first aircraft ever designed for the multiple role of tactical fighter.

It did not take long for us to decide on the Phantom. For one thing it had most of the records in the book. While that may seem colorful to some, to the tactical fighter pilot performance equals survival. He needs an aircraft capable of meeting the enemy on his own terms, at the enemy's speed or better and at his altitude. For the interdiction role he needs an aircraft with a great weight carrying capability. For the close support role he needs a steady gun platform, one he can slow enough to acquire his target and then have the rapid acceleration necessary to escape return ground fire, both from hand held and radar controlled weapons.

This amounts to a pretty difficult problem for the designer, the manufacturer and the services. We have to give a little and take a little down the line remembering all the while the most important thing is survivability of the weapon and pilot for tomorrow's mission.

These were some of the reasons leading up to the announcement in January of last year that we would purchase both reconnaissance and tactical fighter versions of the Phantom. Only a few days after that announcement the Navy loaned two of its fighters to the Air Force, complete with USAF insignia for our test purposes. Although we had a very good idea how they operated we wanted to try them with USAF air and ground crews. We conducted tests on our gunnery ranges and built up a backlog of experience with the aircraft. We found it capable of performing all tactical roles although its primary mission had been air superiority.

We are modifying the radar operator's seat to that of co-pilot as a bonus feature for the Phantom. One of the cardinal advantages of tactical airpower is its mobility. A squadron in the US today can operate in Southeast Asia or the Middle East tomorrow. In fact it must be able to. Otherwise we would find ourselves in the position of purchasing more and more weapons that can perform fewer and fewer functions.

We made minor and major modifications to allow a more practical wheel and tire for operations in rough landing areas requiring a wider flexibility in a variety of conditions.

The role the Navy has played in getting this aircraft into the USAF inventory deserves further mention. This winter we sent a number of ground personnel and instructor pilots to their schools at Oceana NAS. These are the people who will now form the initial group of personnel in the Combat Crew Training Squadron at MacDill AFB. Throughout the entire program the Navy has been extremely helpful.

It can be stated that the combined efforts of the Air Force and Navy in this program will give Tactical Air Command one of the most versatile tactical fighters in existence for the next few years.

Moreover, the utilization of this multi-service flexible fighter by the Navy, Marines, and Air Force will result in considerable savings to the taxpayers, something we are all interested in.[2]

The USN's F4H-1s were eventually re-designated as F-4As in 1962. These aircraft featured J79-GE-2 and -2A engines as powerplants. Later variants operated with -8 engines. On 25 March 1961, the

Lieutenant General Gabriel P. Disosway, USAF Directorate of Programs and Requirements, displays an F-4C model at a Pentagon press conference on 5 February 1963. (US National Archives at College Park, Maryland, Still Pictures Branch)

An F-4C, armed with a variety of air-to-ground weapons, awaits weapons flight tests on the flight line at Nellis AFB, Nevada, in March 1962. (US National Archives at College Park, Maryland, Still Pictures Branch)

An F-4C, loaded up with 750lb bombs, heads out to perform a fighter-bomber feasibility test at Nellis AFB, Nevada, in March 1962. (US National Archives at College Park, Maryland, Still Pictures Branch)

An F-4C, loaded up with M-117 bombs and underwing pylon fuel tanks, awaits flight-testing at Edwards AFB, California, on 3 June 1964. (US National Archives at College Park, Maryland, Still Pictures Branch)

primary Phantom II destined for USN and Marine Corps service, the F-4B, successfully performed its maiden flight. The F-4B featured a Westinghouse APQ-72 radar (pulse only); Texas Instruments AAA-4 infrared seeking and tracking pod (located beneath the nose); AN/AJB-3 bombing system; and J79-GE-8, -8A and -8B engine powerplants complete with afterburners. Typical armament of the F-4B consisted of four AIM-7 Sparrow radar-guided medium-range air-to-air missiles and four AIM-9 Sidewinder heat-seeking short-range air-to-air missiles for air superiority missions, one centre-line external fuel tank, and two outboard external fuel tanks. The F-4B could also carry a variety of gravity bombs (including anti-personnel cluster bombs), napalm bombs, Bullpup air-to-ground missiles, or rocket pods for ground-attack missions. A total of 649 F-4Bs were produced. The first F-4Bs were delivered to NAS Miramar's VF-121 Pacemakers in 1961.

The navy's and marine corps' next version of the Phantom II, the F-4J, featured enhanced aerial combat and bomber qualities. F-4Js began to roll off the assembly lines in 1966 with the final aircraft variant produced being delivered in 1972. A total of 522 F-4Js were produced. The aircraft featured J79-GE-10 engines capable of generating 17,844lbf thrust; a Westinghouse AN/AWG-10 Fire Control System, which provided

the F-4J with look-down/shoot-down capability; and an enhanced integrated missile control system as well as the enhanced AN/AJB-7 bombing system. Thus, the F-4J had become the ultimate fighter-bomber of its time. Armament carried by the F-4J was similar to that carried by the F-4B.

A view inside the pilot's cockpit and of the instrument panels in an F-4 Phantom II. (Beethoven)

In 1972, the USN commenced Project Bee Line, an effort to further develop the successful F-4B design into the F-4N. The F-4N featured smokeless engines as well as F-4J aerodynamic refinements. Approximately 228 F-4Bs were converted to F-4Ns by 1978. An even more advanced version of the Phantom II, designated the F-4S, also later served both the USN and USMC. This variant emerged during the late 1970s from approximately 265 modified F-4Js. The F-4S featured J79-GE-17 smokeless engines (capable of 17,900lbf thrust), AWG-10B radar, Honeywell AN/AVG-8 Visual Target Acquisition Set or VTAS, avionics enhancements and the addition of leading edge slats that significantly improved aircraft maneuverability. Armament carried by F-4Ns and F-4Ss were similar to that carried by the F-4J and F-4B.

The RF-4B photoreconnaissance Phantom II variant was provided to the USMC in March 1965. A fleet of approximately 46 RF-4Bs were in service with the USMC. The RF-4B featured J79-GE-8 engines, navy IFR systems, aircraft controls in the pilot's cockpit position only, and in-flight rotating KS-72 or KS-85 cameras.

The RF-4C photoreconnaissance Phantom II entered service with the USAF on 8 August 1963. A modified F-4B served as the prototype aircraft. The RF-4C had many of the same features as the F-4C, including similar engines and landing gear. New on the aircraft, however, was a lengthened nose, which was extended by an additional 4ft 8in to accommodate either KS-72 or KS-77 forward-framing oblique cameras, a KA-55A high-altitude panoramic camera, and a KA-56A low-altitude panoramic camera. The aircraft also featured an APQ-99 forward-looking radar complete with terrain guidance features as well as an APQ-102 Side Looking Radar (SLR).

On 9 December 1965, the USAF received its first advanced version of the F-4C, the F-4D. The aircraft was essentially similar to the F-4C, but featured enhanced avionics and ground-attack capability. It was equipped with an APQ-109 partial solid-state radar, ASG-22 Lead Computing Optical Sight System (LCOSS), and ASQ-91 weapons computer. In addition to the same array of air-to-air armament carried by the other Phantom IIs, the F-4D could carry 18,750lb of gravity bombs, 15,680lb of mines, eleven 1,000lb bombs, eleven 150 US gallon napalm bombs, four Bullpup air-to-ground missiles, or three 20mm Vulcan gun pods.

On 7 August 1965, the F-4E entered service with the USAF. F-4E development occurred at the same time as the navy's F-4J was developed. While it was hoped that the F-4D could be further upgraded, enhancements to the F-4's radar systems – namely the incorporation of

An F4H model mounted in the Unitary Plan Wind Tunnel for tests at NASA Langley Research Center in Hampton, Virginia. (NASA Langley Research Center)

the small APQ-120 pulse radar and the USAF's need and desire for an F-4 equipped with an internal cannon for close-in aerial combat with North Vietnamese Air Force MiGs – led to the development of the F-4E. The F-4E was equipped with an M61-A1 Vulcan cannon and modified (extended) nose, wing leading-edge slats for improved maneuverability, slotted leading-edge stabilisers, an APQ-120 pulse Doppler radar/ intercept computer, and J79-GE-17 engines capable of providing 17,900lbf thrust.[3]

On 6 December 1975, the USAF received its first F-4G Wild Weasel, an F-4 variant similar to the F-4E but specially equipped to perform 'Wild Weasel' anti-SAM site missions. The F-4G was equipped with a specially designed chin pod (the Vulcan cannon was removed) that concealed an APR-38 Radar Homing And Warning System (RHAWS) threat detecting antennae system and an additional pod atop the tail fin. Typical armament carried by the F-4G consisted of Shrike and Standard anti-radiation missiles, High-Speed Anti-Radiation Missiles (HARMs), or Maverick anti-armour air-to-ground missiles.

A total of 5,057 Phantom IIs were produced in the United States, with aircraft production ceasing in 1979.

NASA's Contributions to Phantom II Development

NASA's contributions to the development of the Phantom II were manyfold. A slight application of famous NASA Langley Research Center engineer Richard T. Whitcomb's area rule concept (a 'pinched-in' or 'wasp-waist' fuselage) was incorporated in the aircraft's design to enable it to fly at supersonic speeds more easily. Wind-tunnel research performed in the NASA Langley Unitary Plan Wind Tunnel indicated that the original F-4 design suffered from lateral-directional stability issues when flying at supersonic speeds. Consequently, McDonnell reworked the wing and tail designs. NASA Langley also studied the stall departure and flat spin characteristics of the Phantom II, and determined the sources of such behavior. Through co-operative research, NASA Langley, McDonnell Douglas, the air force, and the navy devised and developed the wing leading-edge slats on later Phantom II variants that significantly enhanced their maneuverability. NASA Langley research also led to the solution of an extreme nose buffet issue with the RF-4B and RF-4C variants.[4]

An F-4E model mounted in the NASA Langley 7 × 10-Foot High-Speed Tunnel awaits testing to analyse and evaluate the effects of wing leading-edge modifications on the aircraft's maneuverability characteristics. (NASA Langley Research Center)

Langley researcher Sue Grafton poses with a slatted F-4 free-flight model that was tested in the NASA Langley Full-Scale Tunnel. (NASA Langley Research Center)

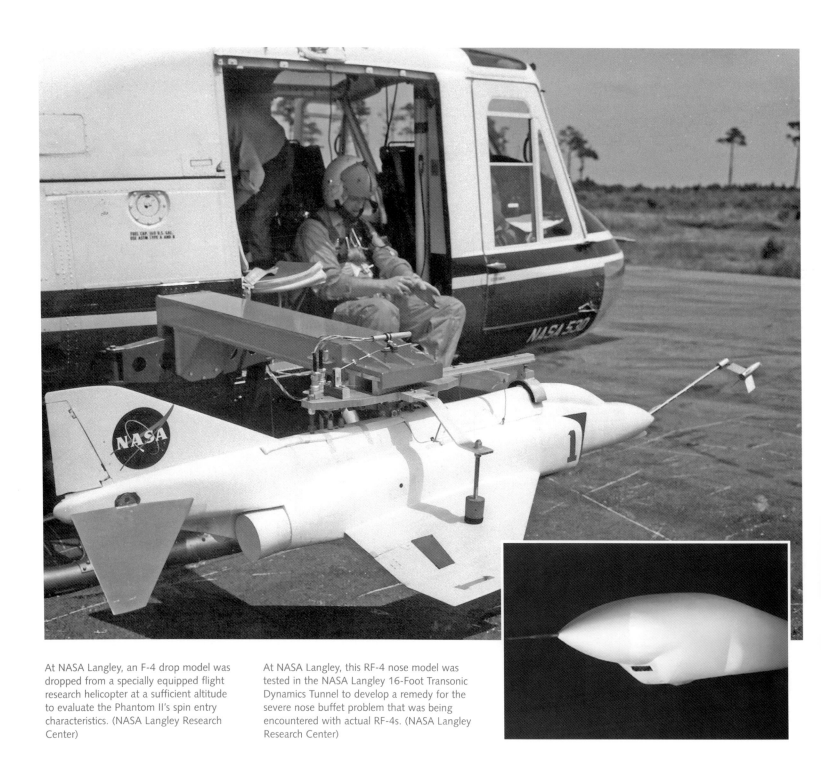

At NASA Langley, an F-4 drop model was dropped from a specially equipped flight research helicopter at a sufficient altitude to evaluate the Phantom II's spin entry characteristics. (NASA Langley Research Center)

At NASA Langley, this RF-4 nose model was tested in the NASA Langley 16-Foot Transonic Dynamics Tunnel to develop a remedy for the severe nose buffet problem that was being encountered with actual RF-4s. (NASA Langley Research Center)

THE PHANTOM II ENTERS OPERATIONAL SERVICE

The Phantom II first became operational with the US Navy's VF-121 Pacemakers at NAS Miramar on 30 December 1960. The unit flew F4H-1Fs or F-4As. NAS Oceana's VF-74 Be-devilers gained the distinction of flying the first combat-operational Phantom IIs with their F4H-1s, also known as F-4Bs, beginning on 8 July 1961. Following the successful completion of carrier trials in October 1961, the unit successfully completed the aircraft type's initial carrier deployment from August 1962 through March 1963, which was performed aboard USS *Forrestal*. The next combat-operational US Navy F-4B unit was the VF-102 Diamondbacks, serving aboard USS *Enterprise*. The VF-114 Aardvarks became the first US Pacific Fleet unit to operate combat-operational F-4Bs. VF-114 F-4Bs were deployed aboard USS *Kitty Hawk* in September 1962. In June 1962, the first combat-operational F-4Bs began service with the US Marine Corps' VMFA-314 (Black Knights) at Marine Corps Air Station El Toro, California. The USAF received their first operational F-4 Phantom IIs under the designation F-110A Spectre in 1963. The F-110As were later designated F-4Cs.

VF-74 became the first combat-operational Phantom II squadron in the US Navy in 1961. (US Navy)

PHANTOM IIs DURING THE CUBAN MISSILE CRISIS

When the Cuban Missile Crisis of late October 1962 escalated, US Navy F-4B Phantom IIs based aboard USS *Enterprise* (CVAN-65) and USS *Independence* (CVA-62) were charged with providing fleet defence and air superiority duties for the US naval blockade of Cuba. In addition, NAS Oceana-based VF-41 F-4Bs were deployed to NAS Key West, Florida to strike targets in Cuba should tensions climax. Although they never saw combat during the crisis, as the crisis was peacefully resolved through diplomatic negotiations, the USN Phantom IIs served as valuable deterrents to any Soviet aerial or ground-based threat to both the US naval fleet and the United States.

PHANTOM IIs OVER VIETNAM

US Navy Phantom IIs in Vietnam

United States Navy Phantom IIs received their baptism of fire during the Vietnam War on 6 August 1964, when they took part in a reprisal attack mission against North Vietnamese gun boat bases in response to the Gulf of Tonkin incident, in which US destroyers were attacked by North Vietnamese gun boats. A total of five USS *Constellation* based VF-142 and VF-143 F-4Bs participated in the mission. By April 1965, President Johnson had ordered a 'limited' bombing offensive against North Vietnam and a full-fledged war erupted in Vietnam. USAF and USMC Phantom IIs also joined the battle.

US Navy, USMC and USAF Phantom IIs performed countless close-air support combat sorties during the Vietnam War, bombarding Viet Cong positions and forces with cluster anti-personnel bombs. In addition, they hammered North Vietnamese regular Army (NVA) positions with millions of pounds of unguided munitions, including the deadly high-drag 'Snake Eye' and 'Fuse Extender' bombs. Napalm bombs were also a choice weapon to suppress enemy ground forces that threatened to overrun US Army and Marine Corps ground positions.

While performing combat air patrol duty on 17 June 1965, US Navy VF-21 F-4Bs shot down two North Vietnamese Air Force MiG-17s, using radar-guided Sparrow air-to-air missiles. On 13 July 1966, a USS *Constellation*-based VF-161 F-4B, piloted by Lieutenant William M. McGuigan and his RIO Lieutenant (junior grade) Robert M. Fowler, shot down a NVAF MiG-17 using an AIM-9D Sidewinder heat-seeking air-to-air missile. On 24 April 1967, a USS *Kitty Hawk*-based VF-114 F-4B, flown by Lieutenant Commander Charles E. Southwick and his RIO Ens. James W. Laing, downed a NVAF MiG-17 with an AIM-9B Sidewinder while providing MiG Combat Air Patrol (MiGCAP) escort for attack aircraft attacking the Kep Airfield. Southwick's and Laing's aircraft was hit by North Vietnamese anti-aircraft fire during the mission, and on the return trip back to *Kitty Hawk* had to eject from their stricken aircraft. Fortunately, they both were later rescued. On 10 July 1968, a USS *America*-based VF-33 F-4J, flown by pilot Lieutenant Roy Cash, Jr, and his RIO Lieutenant Joseph E. Kain, Jr, downed a NVAF MiG-21 using an AIM-9D Sidewinder heat-seeking air-to-air missile while performing MiGCAP duty. A USS *Constellation*-based VF-142 F-4J, flown by Lieutenant Jerome E. Beaulier and his RIO Lieutenant Steven J. Barkley, shot down a NVAF MiG-21 using AIM-9D Sidewinder missiles on 28 March 1970. A USS *Coral Sea*-based VF-111 F-4B, flown by Lieutenant Garry L. Weigand and his RIO Lieutenant (junior grade) William C. Freckelton, downed a NVAF MiG-17 using an AIM-9G Sidewinder missile on 6 March while providing fighter cover for an aerial recon aircraft surveilling Wuang Lang airfield. On 6 May 1972, a USS *Kitty Hawk*-based VF-114 F-4J, flown by Lieutenant Commander Kenneth W. Pettigrew and his RIO Lieutenant (junior grade) Michael J. McCabe, downed a NVAF MiG-21 using an AIM-9G Sidewinder while flying MiGCAP for a follow-up airstrike on the Bia Thuong airfield.

VF-84 F-4Bs provided fleet defence and MiGCAP for USS *Independence* operating in the Gulf of Tonkin in 1965. (US Navy)

A USS *Midway*-based F-4B en route to strike a Viet Cong position in September 1965. (US National Archives at College Park, Maryland, Still Pictures Branch)

F-4 Phantom IIs are catapult-launched from the nuclear-powered USS *Enterprise*, stationed in the Gulf of Tonkin, to perform combat sorties above North Vietnam on 5 April 1966. (US National Archives at College Park, Maryland, Still Pictures Branch)

An F-4B Phantom II returns to the USS *Enterprise* after striking North Vietnam on 3 April 1966. (US National Archives at College Park, Maryland, Still Pictures Branch)

A major milestone of the Vietnam War was later achieved on 10 May 1972, when a Navy F-4J (Showtime 100), flown by Lieutenant Randy 'Duke' Cunningham and his RIO Lieutenant (junior grade) William P. Driscoll, downed three North Vietnamese MiG-17s using AIM-9G Sidewinder missiles, making the Phantom's aircrew the first American aerial aces of the conflict. Another monumental achievement was attained during the deadly engagement–their fifth kill was a MiG-17 piloted by none other than North Vietnamese ace Colonel Nguyen Toon. The Phantom's ruggedness and remarkable ability to withstand substantial punishment was clearly evidenced during Showtime 100's lethal sortie, when the aircraft sustained significant damage from a North Vietnamese SAM and its aircrew piloted their heavily damaged aircraft over the South

A USS *Constellation*-based VF-161 F-4B undergoes aerial refuelling by a Heavy Attack Squadron 8 A-3B Skywarrior on 17 September 1966. The *Constellation* was operating in the South China Sea at the time. (US National Archives at College Park, Maryland, Still Pictures Branch)

China Sea to eject. US Navy F-4Bs, F-4Js and F-4Ns performed a total of eighty-four combat tours during the conflict in Vietnam. A total of forty aerial victories were registered by US Navy Phantom IIs, while incurring a loss of seventy-three of their own type. Of these seventy-three losses, only seven were lost to hostile aircraft, thirteen were lost to SAMs and fifty-three were lost to AAA.[5]

An F-4J Phantom II following a catapulted launch from USS *America*, operating in the South China Sea on 26 May 1968. (US National Archives at College Park, Maryland, Still Pictures Branch)

A USS *Coral Sea*-based VF-111 Sundowners F-4B performs a bombing run above Vietnam in 1971. (US Navy)

A VF-96 F-4J Phantom II (left) and a Reconnaissance Attack Squadron 5 (RVAH-5) RA-5C Vigilante, based aboard USS *Constellation*, perform a flight in support of Operation Midlink '74 on 17 November 1974. (US National Archives at College Park, Maryland, Still Pictures Branch)

A pair of VF-96 F-4Js commence landing preparations for a landing aboard USS *Constellation* on 28 November 1974. The aircraft just completed a mission in support of Operation Midlink '74. (US National Archives at College Park, Maryland, Still Pictures Branch)

The first American aces of the Vietnam War, pilot Lieutenant Randall H.
Cunningham (right) and Radar Intercept Officer Lieutenant (junior grade) William
P. Driscoll, in the cockpit of their F-4J Phantom II on 10 May 1972. (US National
Archives at College Park, Maryland, Still Pictures Branch)

A pair of VF-151 F-4Bs make a bomb run on a North Vietnamese target, while a USAF F-4D serves as a guide aircraft. (US National Archives at College Park, Maryland, Still Pictures Branch)

A trio of USS *Midway*-based VF-161 F-4Bs and a trio of USS *America*-based A-7E Corsair IIs perform a Long-Range Navigation bombing run on a North Vietnamese target in 1973. (US National Archives at College Park, Maryland, Still Pictures Branch)

USMC Phantom IIs in Vietnam

F-4Bs, assigned to the VMFA-531 Gray Ghosts, became the first USMC Phantom IIs to be deployed to the Vietnam War and were based at Da Nang airbase beginning on 10 May 1965. Although their first role upon arrival was to perform air-defence duty, the USMC Phantom IIs later performed close air-support sorties. The Gray Ghosts were supported in this role by the arrival of the VMFA-314 Black Knights, VMFA-232 Red Devils, VMFA-323 Death Rattlers and VMFA-542 Bengals. USMC F-4Bs also proved to be effective in aerial combat with the North Vietnamese Air Force, downing one MiG-21 and USMC F-4B aircrews, flying USAF Phantom IIs as part of an exchange programme, downing two more MiGs. The USMC F-4B that scored a victory over the NVAF MiG-21 was flown by Major John P. Hefferman (USAF) and his RIO Lieutenant (junior grade) Frank A. Schumacher. These men registered their aerial victory on 9 May 1968 via use of an AIM-7E Sparrow radar-guided missile. A total of seventy-five USMC Phantom IIs were shot down during the war. The majority of these losses were attributable to enemy ground fire, while four were lost in accidents.

The Marine Corps also operated the RF-4B photo reconnaissance Phantom II in the skies over Vietnam. The Da Nang-based VMCJ-1 Golden Hawks performed the initial RF-4B photo reconnaissance sortie of the war on 3 November 1966 and maintained operations from Da Nang until 1970. The squadron suffered no losses, with only one aircraft ever being damaged, attributed to enemy AAA. USMC Phantom IIs remained in service from the 1970s to the early 1990s.

A Marine Fighter/Attack Squadron 323 Death Rattlers F-4B awaits action on the flight line at Chui Lai. The aircraft is loaded up with napalm bombs and is well equipped to perform close air support. (US National Archives at College Park, Maryland, Still Pictures Branch)

Da Nang-based VMFA-542 F-4Bs en route to provide close air support for Northern I Corps Marines. (US National Archives at College Park, Maryland, Still Pictures Branch)

A bombed-up USMC F-4B at rest in a revetment at MAG-13, Chui Lai, in December 1967. (US National Archives at College Park, Maryland, Still Pictures Branch)

A pair of Da Nang-based VMFA-542 F-4Bs en route to provide close air support for Northern I Corps Marines. The aircraft are loaded up with 'Snakeye' high-drag bombs. (US National Archives at College Park, Maryland, Still Pictures Branch)

A USMC RF-4B (right) and a USMC F-4B performing a mission. (US National Archives at College Park, Maryland, Still Pictures Branch)

Marine Composite Reconnaissance Squadron 1 (VMCJ-1) ground crews greet an RF-4B air crew upon the aircraft's return to base following an aerial photoreconnaissance mission on 30 January 1970. (US National Archives at College Park, Maryland, Still Pictures Branch)

A Marine Composite Reconnaissance Squadron 1 (VMCJ-1) RF-4B, with drag chute deployed, lands at its base following completion of an aerial photoreconnaissance mission on 30 January 1970. (US National Archives at College Park, Maryland, Still Pictures Branch)

A USAF F-4C Phantom II on a photo chase sortie above Vietnam in October 1965. The aircraft is carrying type IV camera pods on the inboard wing pylons and four Sparrow radar-guided air-to-air missiles. (US National Archives at College Park, Maryland, Still Pictures Branch)

USAF Phantom IIs in Vietnam

The first USAF Phantom II squadron to be deployed to Vietnam was the 555th Tactical Fighter Squadron 'Triple Nickel' in December 1964, which operated F-4Cs. The 45th Tactical Fighter Squadron, 15th Tactical Fighter Wing F-4Cs, achieved another milestone of the Vietnam War when they registered the first two USAF MiG-17 kills of the war on 10 July 1965. The two USAF Phantom IIs, operating from Ubon, Thailand, used AIM-9B Sidewinder heat-seeking air-to-air missiles to down their prey. Another first was achieved on 26 April 1966, when a 480th Tactical Fighter Squadron F-4C, flown by Major Paul J. Gilmore and his Weapons System Officer (WSO) 1st Lieutenant William T. Smith, downed the first North Vietnamese MiG-21 Fishbed to be shot down by a US aircraft using an AIM 9-B Sidewinder missile. On 26 April 1967, a 389th Tactical Fighter Squadron – 366th Tactical Fighter Wing F-4C, flown by Major Rolland W. Moore, Jr, and his WSO 1st Lieutenant James F. Sears – shot down a NVAF MiG-21 using an AIM-7E Sparrow radar-guided air-to-air missile while escorting a formation of F-105 Thunderchiefs en route to strike a transformer facility in Hanoi. On 14 May 1967, a 480th Tactical Fighter Squadron – 366th Tactical Fighter Wing F-4C, flown by Major James A. Hargrove, Jr, and his WSO 1st Lieutenant Stephen H. De Muth – shot down a North Vietnamese MiG-17 using a 20mm SUU-16 gun pod. Major Hargrove and 1st Lieutenant De Muth were escorting a formation of F-105 Thunderchiefs performing a ground-attack mission on the Ha Dong army barracks and supply depot.

One of the most successful USAF Phantom II pilots during the early portion of the Vietnam War was Second World War ace Colonel Robin Olds, who became Wing Commander of the 8th Tactical Fighter Wing, stationed at Ubon Royal Thai AFB, Thailand, on 30 September 1966. During his tour in the Vietnam War, Olds piloted an F-4C Phantom II, dubbed 'Scat XXVII', in which he shot down two North Vietnamese Air Force MiG-21 Fishbeds and two NVAF MiG-17 Frescos. Flying as F-105 Thunderchief decoys in Operation Bolo, 8th Tactical Fighter Wing F-4Cs destroyed seven MiG-21s, with Olds and his WSO downing one of them on 2 January 1967. Olds and his WSO were later presented with Silver Stars for their accomplishment. Olds downed his second MiG-21 in the skies above the MiG-21 base at Phúc Yên on 4 May 1967, and on 20 May 1967 he downed two MiG-17s. These victories brought his Vietnam War kill tally to four. Having already registered twelve aerial victories during the Second World War, his Vietnam War victory tally earned him his new title – 'triple ace'.

In December 1966, Colonel Olds was joined in command of the 8th Tactical Fighter Wing by Colonel Daniel 'Chappie' James, Jr, who became 8th Tactical Fighter Wing Vice Commander one year later. James successfully performed seventy-eight sorties over North Vietnam. Most notably, he commanded a 'Bolo' MiG sweep mission that resulted in the downing of seven North Vietnamese MiG-21s, a kill tally that was to be the largest, for a single mission, of the Vietnam War. James and Olds were often affectionately referred to as 'Blackman and Robin' within USAF units.

Later during the Vietnam War, USAF F-4D aircrews began amassing impressive victory tallies over NVAF MiGs. On 6 February 1968, a 433rd Tactical Fighter Squadron – 8th Tactical Fighter Wing F-4D, flown by Captain Robert H. Boles and his WSO 1st Lieutenant Robert B. Battista – shot down a NVAF MiG-21 using an AIM-7E Sparrow missile. On 28 August 1972, Udorn Royal Thai AFB-based 555th ('Triple Nickel') Tactical Fighter Squadron F-4D pilot Captain Steve Ritchie downed his fifth NVAF MiG-21, making him the first and sole USAF ace of the Vietnam War. Ritchie registered MiG-21 kills previously on 10 May, 31 May and 8 July 1972. Ritchie managed to score all five of his MiG-21 victories using the AIM-7 Sparrow radar-guided air-to-air missile.

Another aerial combat first of the Vietnam War was achieved on 2 June 1972, when a 58th Tactical Fighter Squadron – 432 Tactical Fighter Wing F-4E, piloted by Major Phillip W. Handley and his WSO 1st Lieutenant John J. Smallwood – registered the first supersonic gun aerial victory of the Vietnam War. The feat was accomplished using the F-4E's internal M61A1, 20mm cannon on a NVAF MiG-19 Farmer in the skies above Vietnam's Thud Ridge at a speed of Mach 1.2.[6, 7]

During the Vietnam War, USAF Phantom IIs proved to be highly effective bombers, often eliminating ground targets with precision and providing critical close air support for troops on the ground. A first in the history of aerial warfare was achieved by the Thailand-based 8th Tactical Fighter Wing F-4Ds in May 1968, when they became the first Tactical Fighter Squadron to use laser-guided bombs in battle. On 10 May 1972, 435th Tactical Fighter Squadron – 8th Tactical Fighter Wing F-4Ds – destroyed the Paul Doumer bridge, located near Hanoi, using Mk.84 laser-guided bombs. Laser-guided bombs were also dropped on targets in North Vietnam by USAF F-4Es. Also on 10 May 1972, other USAF

F-4s struck the Yen Vien railway yard and numerous communications sites within close proximity using 500lb iron bombs.

A total of four squadrons flew RF-4Cs during the Vietnam War, eighty-three of which were lost. Seventy-two of these eighty-three losses were attributed to enemy ground fire – seven SAM casualties and sixty-five AAA casualties.[8]

During the war, USAF F-4C/D/Es amassed a tally of 107½ MiGs destroyed – fifty using Sparrow missiles, thirty-one using Sidewinder missiles, five using Falcon missiles, fifteen and a half using guns and six using other methods.[9] Approximately 370 USAF Phantom II fighter-bombers fell victim to enemy fire during the conflict in Vietnam – thirty-three lost to gun and missile fire from MiGs, thirty SAM casualties and 307 AAA casualties.[10]

USAF F-4Cs on a strike mission in Vietnam on 6 November 1965. (US National Archives at College Park, Maryland, Still Pictures Branch)

An F-4C Phantom II, with drag chute deployed, completes its landing at operating location Ol Cocoa in Vietnam in November 1965. (US National Archives at College Park, Maryland, Still Pictures Branch)

A USAF F-4C performs a landing at Ol Cocoa, Republic of Vietnam, on 9 December 1965. (US National Archives at College Park, Maryland, Still Pictures Branch)

A USAF 431st Tactical Fighter Squadron F-4C, sporting a camouflage scheme and loaded up with eight 750lb general-purpose bombs and two camera pods, awaits action at Ol Cocoa, Republic of Vietnam, on 9 December 1965. (US National Archives at College Park, Maryland, Still Pictures Branch)

USAF F-4Cs on a mission above Vietnam in 1965. (US National Archives at College Park, Maryland, Still Pictures Branch)

A USAF F-4C above South Vietnam in 1965. (US National Archives at College Park, Maryland, Still Pictures Branch)

A USAF F-4C taxis past a Search and Rescue rotorcraft on the flight line at an airbase in South Vietnam in 1965. (US National Archives at College Park, Maryland, Still Pictures Branch)

Four USAF F-4Cs on their way back to base following a strike mission above North Vietnam in November 1965. (US National Archives at College Park, Maryland, Still Pictures Branch)

A USAF 43rd Tactical Fighter Squadron F-4C embarks on a close-air support sortie from Cam Ranh airbase, Vietnam, in November 1965. (US National Archives at College Park, Maryland, Still Pictures Branch)

A bombed-up USAF F-4C above Vietnam in 1965. (US National Archives at College Park, Maryland, Still Pictures Branch)

Three USAF F-4Cs streak over the skies of Vietnam in 1965. (US National Archives at College Park, Maryland, Still Pictures Branch)

Four F-4Cs, armed with bombs, on their way to strike a North Vietnamese target in Vietnam in 1965. (US National Archives at College Park, Maryland, Still Pictures Branch)

A USAF F-4C, loaded with bombs, en route to strike an enemy position in Vietnam in 1965. (US National Archives at College Park, Maryland, Still Pictures Branch)

Four F-4Cs, loaded up with 750lb bombs, above Vietnam in December 1965. (US National Archives at College Park, Maryland, Still Pictures Branch)

USAF F-4Cs above Vietnam in 1965. (US National Archives at College Park, Maryland, Still Pictures Branch)

A USAF F-4C, armed with camera pods and a 20mm gun pod, heads out from its base in South Vietnam on a close-air support sortie on 12 February 1966. (US National Archives at College Park, Maryland, Still Pictures Branch)

A camouflaged USAF F-4C, bristling with armament, performs a sortie over Vietnam in December 1965. (US National Archives at College Park, Maryland, Still Pictures Branch)

A USAF F-4C Phantom II passes a guard post at Da Nang airbase, Vietnam, in 1966. (US National Archives at College Park, Maryland, Still Pictures Branch)

A USAF F-4C Phantom II comes in for a landing at Da Nang airbase, South Vietnam, in June 1966. (US National Archives at College Park, Maryland, Still Pictures Branch)

A bomb-laden F-4C heads out on another combat sortie from Da Nang airbase in October 1966. (US National Archives at College Park, Maryland, Still Pictures Branch)

The all-weather operational capability of the F-4 Phantom II is evident in this image of a Da Nang-based USAF 480th Tactical Fighter Squadron F-4C being prepped for a bombing sortie above North Vietnam in November 1966. The aircraft is armed with Snakeye high-drag bombs and a gun pod. (US National Archives at College Park, Maryland, Still Pictures Branch)

USAF Colonel Robin Olds poses beside his F-4C Phantom II, nicknamed 'SCAT XXVII', for a publicity photo. His aircraft was sporting two North Vietnamese MiG kills at the time. (USAF)

A USAF F-4C in flight over the South China Sea. The aircraft just completed a ground-attack sortie over South Vietnam and is returning to Cam Ranh Bay airbase. (US National Archives at College Park, Maryland, Still Pictures Branch)

USAF Colonel Daniel 'Chappie' James, Jr. poses in front of his F-4C Phantom II for a publicity photo at Ubon Royal Thai AFB, Thailand. (USAF)

Five USAF F-4Ds form up in the skies above Thailand on 1 November 1967. (US National Archives at College Park, Maryland, Still Pictures Branch)

The Colour Guard presents the colours in front of a USAF F-4D at Ubon airbase, Thailand, in 1969. (US National Archives at College Park, Maryland, Still Pictures Branch)

A USAF F-4D taxis down the runway, with drag chute open, following landing at Udorn airbase, Thailand, in August 1969. The aircraft was nicknamed 'Mr Snoopy'. (US National Archives at College Park, Maryland, Still Pictures Branch)

A USAF F-4D in flight over Thailand on 23 October 1970. (US National Archives at College Park, Maryland, Still Pictures Branch)

Captain Richard S. ('Steve') Ritchie stands next to his F-4D Phantom II after becoming the first USAF ace of the Vietnam War. (USAF)

A Collings Foundation F-4D at Selfridge Air National Guard Base (ANGB), Michigan, in 2005. The aircraft wears Steve Ritchie's and Chuck DeBellevue's aircraft's markings. (Jacobst at English Wikipedia, via Wikimedia Commons)

A Thai AF T-28 (left) and USAF 8th Tactical Fighter Wing F-4D in position at the 'Last Chance' prep point at Ubon airbase, Thailand, in September 1972. The Last Chance prep point was a final check area where aircraft weapons were armed and loaded aboard aircraft. Note the F-4D's Long Range Navigation antenna (on the fuselage mid-section of the aircraft) used as a bombing aid, and Sparrow radar-guided air-to-air missiles already loaded onto the jet fighter-bomber. (US National Archives at College Park, Maryland, Still Pictures Branch)

A USAF 432nd Tactical Reconnaissance Fighter Wing F-4D in position on the flight line at Udorn airbase, Thailand, in October 1972. Note the huge USAF C-141 transport to the left (back). (US National Archives at College Park, Maryland, Still Pictures Branch)

A side view of the forward fuselage of a 432nd Tactical Fighter Wing F-4D prior to heading out on a mission from Udorn airbase, Thailand, in October 1972. Note the electronic countermeasures pod on the portside, inboard underwing pylon. (US National Archives at College Park, Maryland, Still Pictures Branch)

View of a 432nd TFW F-4D from under the tail section of another F-4 at Udorn airbase, Thailand, in October 1972. (US National Archives at College Park, Maryland, Still Pictures Branch)

An 8th TFW F-4D, equipped with a Long Range Navigation radar antenna, at Ubon airbase, Thailand, in June 1973. (US National Archives at College Park, Maryland, Still Pictures Branch)

An 8th TFW F-4D at Ubon airbase, Thailand, in September 1972. (US National Archives at College Park, Maryland, Still Pictures Branch)

A line up of 8th TFW F-4Ds at Ubon airbase, Thailand, in September 1972. (US National Archives at College Park, Maryland, Still Pictures Branch)

One of the Phantom II's many strengths was its ability to take harsh punishment and bring its aircrews safely back home. This USAF F-4 withstood a nearby surface-to-air missile explosion during a combat sortie in 1973. (US National Archives at College Park, Maryland, Still Pictures Branch)

A six-plane formation of eighteen USAF F-4Es arrive at Korat airbase, Thailand, on 17 November 1968.
This was the F-4E's maiden deployment to the Vietnam War and the aircraft were detached to the 40th
Tactical Fighter Squadron. (US National Archives at College Park, Maryland, Still Pictures Branch)

The initial Vietnam War combat-operational F-4E Phantom IIs taxi in at Korat airbase, Thailand, on 17 November 1968. (US National Archives at College Park, Maryland, Still Pictures Branch)

A USAF F-4E lands at Korat airbase, Thailand, on 17 November 1968. (US National Archives at College Park, Maryland, Still Pictures Branch)

The first USAF F-4Es to be deployed to the Vietnam War at Korat airbase,
Thailand, following their arrival on 17 November 1968. (US National Archives at
College Park, Maryland, Still Pictures Branch)

A USAF F-4E, nicknamed 'Little Chris', taxis at Korat airbase, Thailand, in 1970. (US National Archives at College Park, Maryland, Still Pictures Branch)

A USAF 432nd Tactical Reconnaissance Fighter Wing F-4E, with drag chute open, lands at Udorn airbase, Thailand, in October 1972. (US National Archives at College Park, Maryland, Still Pictures Branch)

A USAF F-4D, loaded up with snakeye high-drag bombs and zuni air-to-ground rockets, awaits action at Ubon airbase, Thailand in 1967. (US National Archives at College Park, Maryland, Still Pictures Branch)

A USAF F-4D, loaded up with Mk.82 snakeye high-drag bombs, an electronic countermeasure pod and dispenser/cluster bombs, awaits action at Ubon airbase, Thailand, on 4 October 1968. (US National Archives at College Park, Maryland, Still Pictures Branch)

A USAF F-4D, well-armed for a close-air support mission, at Udorn airbase, Thailand, on 20 September 1968. (US National Archives at College Park, Maryland, Still Pictures Branch)

A USAF 432nd TRFW F-4D, loaded up with fuse-extender bombs, takes off from Udorn airbase, Thailand, on a combat sortie above North Vietnam in October 1972. (US National Archives at College Park, Maryland, Still Pictures Branch)

A USAF F-4E is prepped for a ground-attack mission at Korat airbase, Thailand, in 1968. (US National Archives at College Park, Maryland, Still Pictures Branch)

In November 1970 a USAF 388th TFW F-4E, loaded up with Mk.82 and cluster bombs at Korat airbase, Thailand, awaits action. (US National Archives at College Park, Maryland, Still Pictures Branch)

A USAF 8th TFW F-4E taxies out onto the runway at Ubon airbase, Thailand, in September 1972. The aircraft is heading out to strike ground targets in North Vietnam. (US National Archives at College Park, Maryland, Still Pictures Branch)

An overhead front view of a USAF 433rd TFS, 8th TFW F-4D in a revetment with the three essential parts of a 2,000lb Paveway laser-guided bomb as well as its guidance hardware at Ubon airbase, Thailand, on 16 September 1969. (US National Archives at College Park, Maryland, Still Pictures Branch)

A ground-level front view of the previous image. (US National Archives at College Park, Maryland, Still Pictures Branch)

A USAF F-4D, carrying Paveway I laser-guided bombs, taxies out of a revetment at Ubon airbase, Thailand, on 16 September 1969. (US National Archives at College Park, Maryland, Still Pictures Branch)

Under the guidance of a ground-crewman, a USAF F-4D, carrying 2,000lb Paveway I laser-guided bombs, taxies to the arming area at Ubon airbase, Thailand, on 16 September 1969. (US National Archives at College Park, Maryland, Still Pictures Branch)

A USAF F-4D, carrying 2,000lb Paveway I laser-guided bombs, heads out from Ubon airbase, Thailand, with its afterburners lit to strike targets in North Vietnam on 16 September 1969. (US National Archives at College Park, Maryland, Still Pictures Branch)

A formation of USAF Ubon airbase-based F-4Ds, loaded up with laser-guided bombs, en route to conduct precision strikes on targets in North Vietnam. (USAF)

A USAF F-4D, loaded up with 3,000lb laser-guided bombs and a pave knife laser-targeting pod, is prepped for a combat sortie at an airbase in Thailand on 12 May 1972. (US National Archives at College Park, Maryland, Still Pictures Branch)

A USAF F-4E, carrying laser-guided bombs, is prepped for a combat mission at an airbase in Thailand on 12 May 1972. (US National Archives at College Park, Maryland, Still Pictures Branch)

A USAF F-4E, loaded up with laser-guided bombs, heads out from Udorn airbase, Thailand, on a combat sortie above North Vietnam in October 1972. (US National Archives at College Park, Maryland, Still Pictures Branch)

A line-up of USAF RF-4C reconnaissance Phantom IIs at U-Tapao airbase, Thailand, in April 1967. Note the B-52D Stratofortress landing (centre) and KC-135 tankers and AC-47 gunships in the background. (US National Archives at College Park, Maryland, Still Pictures Branch)

A USAF RF-4C at Udorn airbase, Thailand, on 20 September 1968. (US National Archives at College Park, Maryland, Still Pictures Branch)

A USAF RF-4C taxies in at Udorn airbase, Thailand, in August 1969. (US National Archives at College Park, Maryland, Still Pictures Branch)

A USAF 11th Tactical Reconnaissance Squadron RF-4C takes off from Udorn airbase, Thailand, headed back to the US on 15 November 1970. (US National Archives at College Park, Maryland, Still Pictures Branch)

A USAF 14th Tactical Reconnaissance Squadron RF-4C is started via engine cartridge in a revetment at Udorn airbase, Thailand, in October 1972. (US National Archives at College Park, Maryland, Still Pictures Branch)

BLUE ANGEL AND THUNDERBIRD PHANTOM IIs

Blue Angel Phantom IIs

From 1969 to 1974, the US Navy's premier aerial demonstration team, the Blue Angels, flew the McDonnell Douglas F-4J Phantom II. The F-4J was the first two-seat, twin engine aircraft flown by the Blue Angels at airshows. During the majority of the airshows, the back seats of Blue Angel Phantom IIs, normally occupied by the 'Guy in Back' or GIB, remained unoccupied. The Phantom II also had the distinction of being the only aircraft to serve both the Blue Angels and the USAF's premier aerial demonstration team, the Thunderbirds, at the same time. Like their Thunderbird counterparts, Blue Angel F-4Js had a propensity to emit quite a bit of smoke from their GE J-79 engines. The Blue Angel F-4Js dazzled crowds during a South American tour in 1970, a Far East Tour (Korea, Japan, Taiwan, Guam and the Philippines) in 1971, and a European tour in 1973 (Tehran, Iran, England, France, Spain, Turkey, Greece and Italy). The Phantom IIs also helped the Blue Angels win the US Navy's Meritorious Unit Commendation (1 March 1970–31 December 1971).

Blue Angels F-4J Phantom II, ship no. 1. (US Navy)

The flight crews of the Blue Angel aerial demonstration team proceed to their aircraft before conducting a performance at Naval Air Station Agana on the island of Guam in the Mariana Islands on 25 November 1971. (US National Archives at College Park, Maryland, Still Pictures Branch)

Blue Angel F-4Js make a transcontinental flight in straight-line formation in 1971. (US Navy)

Blue Angel F-4Js during a performance at the Combined Open House and Air Show at Roosevelt Roads, Puerto Rico, on 23 March 1970. (US National Archives at College Park, Maryland, Still Pictures Branch)

Thunderbird Phantom IIs

From 1969 to 1973, the USAF's premier aerial demonstration team, the Thunderbirds, flew the McDonnell Douglas F-4E Phantom II. The Thunderbirds began to fly their new F-4Es at airshows in spring 1969. The aerial demonstration team's transition to the Phantom II proved to be complex, however. As a result of the Phantom IIs Mach 2+ airspeed capability, the paint schemes of the aircraft had to be changed to enable the paint to withstand the rigours of high-speed flight. Consequently, a white paint base was adopted as the base colour applied to all modern Thunderbird aircraft paint schemes. With F-4Es now in Thunderbird service, their aerial demonstrations were more awesome than Thunderbird aerial demonstrations of the past. Their powerful J-79 engines often woke up dazed crowds, producing a thunderous roar that had not been heard at airshows before. At airshows and flight demonstrations, Thunderbird F-4Es performed Diamond, Solo and 'Bomb Burst' maneuvers. Thunderbird F-4Es often flew the following formations during airshows: (1) Diamond, (2) Delta, (3) Stinger, (4) Arrowhead, (5) Line-Abreast, (6) Trail, (7) Echelon and (8) Five Card.

Thunderbird F-4E Phantom IIs fly in formation in 1972. Note that ship No. 4 is missing its tail number. (USAF)

Thunderbirds pilot Len Moon poses next to his F-4E for a publicity photo in 1973. (USAF)

7

ISRAELI AIR FORCE PHANTOM IIs IN THE WAR OF ATTRITION, YOM KIPPUR WAR, AND OVER THE BEKAA VALLEY AND LEBANON

In 1969, US President Richard Nixon authorised Israel's acquisition of 204 F-4Es to help combat Egyptian and Syrian aggression in the Middle East. The Phantom IIs arrived in Israel on 5 September 1969 and first saw combat during the War of Attrition above the Suez Canal. Israeli Air Force (IAF) F-4Es performed their first strike mission on an SA-2 (SAM) battery on 22 October 1969. Then on 11 November 1969, the IAF Phantom IIs registered their first Egyptian MiG-21 kill, which resulted from an engagement between two IAF Phantoms and four Egyptian MiG -21s. On 30 July 1970, a force of 16 MiG-21s, flown by Soviet pilots, attempted to intercept a strike force of IAF Phantoms and Mirages that had penetrated deep into Egyptian territory. In the ensuing combat, a total of five MiGs were shot down with no IAF losses.[11] There were a total

of seven combat-operational IAF Phantom II squadrons (approximately 121 aircraft) during the Yom Kippur War. While a total of thirty-three IAF F-4Es were shot down during the war, mostly due to enemy ground fire, the IAF enjoyed air supremacy over the Egyptian Air Force, largely attributable to the superior aerial combat performance of the Phantom II. This was clearly evidenced by the aerial combat exploits of IAF Colonel Israel Baharav, who downed fourteen Egyptian MiGs while flying an F-4E during the war.[12] IAF F-4Es were instrumental in the elimination of Syrian SAM sites during the 1982 Bekaa Valley conflict as well as in

future conflicts in Lebanon. IAF F-4Es accounted for the destruction of 116½ hostile aircraft during their IAF service careers. Some of the hostile aircraft destroyed were flown by Russian pilots. During the latter 1980s, an upgraded variant of the IAF Phantom II, known as the Kurnass 2000, became operational and participated in strikes on terrorist camps within Lebanon. In 1974, General Dynamics sought to provide for the IAF three modified F-4Es that had been converted into RF-4E photo-reconnaissance variants. These aircraft participated in the 1982 conflict in Lebanon. The IAF's Phantom II fleet was mothballed in 2004.

Opposite: A decommissioned IAF F-4 Phantom Kurnas (Sledgehammer). The aircraft wears three Syrian Air Force MiG kills on the forward fuselage left side. Note the practice air-to-air missiles and laser-guided bombs on the underwing racks and pylons. (Courtesy of Zachi Evenor from Israel, via Wikimedia Commons)

Above: A decommissioned IAF F-4 Phantom Kurnas. The aircraft's cannon has been uncovered and remains exposed for display. (Courtesy of brewbooks from near Seattle, USA, via Wikimedia Commons)

8

PHANTOM IIs DURING THE COLD WAR

McDonnell Douglas F-4 Phantom IIs served as primary air superiority fighters during the majority of the Cold War, until the more advanced Grumman F-14 Tomcat (US Navy), McDonnell Douglas F-15 Eagle (USAF), and McDonnell Douglas F/A-18 Hornet (USMC) entered service. The Grumman F-14 Tomcat entered service with the US Navy in 1974, replacing Phantom IIs as the primary air superiority fighter in numerous carrier-based fighter squadrons. The McDonnell Douglas F-15 Eagle entered service with the USAF in 1975, replacing Phantom IIs as the primary air superiority fighter in numerous land-based fighter squadrons. The McDonnell Douglas F/A-18 Hornet entered service with the USMC in 1984, replacing Phantom IIs as the primary air superiority fighter in numerous Marine Corps fighter squadrons. Before these modern air superiority fighters entered service, Phantom IIs served as the main air superiority fighter for all three of these US military services, routinely intercepting marauding Soviet long-range reconnaissance/bomber aircraft when they flew into US airspace. In addition to performing Soviet bomber interception duty, Phantom IIs served as test platforms for flight testing new aerial weapons.

Phantom IIs served as flight test platforms for such aerial weapons as the AGM-45 Shrike anti-radiation missile used to destroy enemy SAM sites, the Walleye television-guided glide bomb, the AGM-88 Highspeed Anti-Radiation Missile (HARM) also used to destroy enemy SAM sites, AGM-65 Maverick air-to-ground missiles, and large television-guided glide bombs.

During the height of the Cold War, in 1975, a specialised Phantom II variant was produced for the USAF in the form of the F-4G Wild Weasel, which proved to be effective in hunting and destroying enemy SAM sites. The aircraft also proved to be a highly effective weapons-test platform. Most often, F-4Gs carried Shrike and Standard anti-radiation missiles, but they proved to be instrumental in the testing of AGM-65 Maverick air-to-ground missiles and HARM missiles as well.

Following the Vietnam War, the US Navy and US Marine Corps continued to operate F-4Js, F-4Ns and F-4Ss as fighter-bombers. USAF Air National Guard and reserve units also began to transition to F-4 Phantom IIs, beginning with the 170th Tactical Fighter Squadron/183d Tactical Fighter Group (Illinois Air National Guard) on 31 January 1972.

A VF-143 F-4J Phantom II intercepts a Soviet Tupolev Tu-16 Badger bomber over the Pacific Ocean in 1970. (US National Archives at College Park, Maryland, Still Pictures Branch)

A VF-151 F-4B Phantom II
intercepts a Soviet Tupolev
Tu-95D Bear bomber over the
Pacific Ocean on 15 March 1974.
(US National Archives at College
Park, Maryland, Still Pictures
Branch)

Two USAF F-4Es intercept a
Soviet Tu-95 Bear bomber
above international waters
on 28 September 1980. (US
National Archives at College Park,
Maryland, Still Pictures Branch)

An overhead view of the previous image. (US National Archives at College Park, Maryland, Still Pictures Branch)

A USS *Coral Sea*-based VF-51 F-4B Phantom II in flight above the Coronado Bay Bridge, San Diego, California, on 8 May 1971. (US National Archives at College Park, Maryland, Still Pictures Branch)

A US Navy F-4 test aircraft, equipped with a Walleye glide-bomb at Naval Weapons Center, China Lake, California, on 24 June 1974. (US National Archives at College Park, Maryland, Still Pictures Branch)

A USAF SR-71 Blackbird, based at Edwards AFB, California, and an Air Development Squadron 4 (VX-4) F-4J Phantom II in flight above Point Mugu, California, in July 1972. (US National Archives at College Park, Maryland, Still Pictures Branch)

A USAF test F-4D, equipped with a Shrike air-to-ground anti-radiation missile, used to destroy enemy surface-to-air missile sites, in 1974. (US National Archives at College Park, Maryland, Still Pictures Branch)

An A-6 Intruder attack bomber and an F-4B of Marine Fighter/Attack Squadron 115 (VMFA-115) prior to catapult launch from the USS *Enterprise* in the South China Sea on 15 January 1975. (US National Archives at College Park, Maryland, Still Pictures Branch)

An F-4 makes a catapult launch from the USS *Forrestal*, in the Mediterranean Sea, on 2 June 1975. (US National Archives at College Park, Maryland, Still Pictures Branch)

An American Bicentennial-clad Air Test and Evaluation Squadron 4 (VX-4) F-4B at Point Mugu, California, in 1976. (US National Archives at College Park, Maryland, Still Pictures Branch)

A front view of the previously depicted aircraft. (US National Archives at College Park, Maryland, Still Pictures Branch)

A VF-31 Tomcatters F-4J in flight in 1978. (US Navy)

A VF-121 F-4J at NAS Miramar, San Diego, California, in 1974. (US National Archives at College Park, Maryland, Still Pictures Branch)

A USAF flight test F-4G Phantom II Wild Weasel performs a weapons flight test of an AGM-88A High-Speed Anti-Radiation Missile in 1979. (US National Archives at College Park, Maryland, Still Pictures Branch)

A Michigan Air National Guard F-4C Phantom II at Tyndall AFB, Florida, in October 1980. The aircraft was taking part in the William Tell 1980 aerial gunnery competition. (US National Archives at College Park, Maryland, Still Pictures Branch)

A USAF F-4E flight-test aircraft is armed with a GBU-15 glide bomb for a test at White Sands Missile Range on 1 December 1978. (US National Archives at College Park, Maryland, Still Pictures Branch)

USAF ground crews prepare to load an AIM-9 Sidewinder missile on an F-4 during the William Tell '86 competition. (US National Archives at College Park, Maryland, Still Pictures Branch)

A USAF F-4E flies chase during an AGM-109 Tomahawk cruise-missile test on 27 November 1979. (US National Archives at College Park, Maryland, Still Pictures Branch)

A USAF F-4G in flight above the Tonopah Test Range. (US National Archives at College Park, Maryland, Still Pictures Branch)

A USAF F-4E equipped with four AIM-4 Falcon air-to-air missiles in 1980. (US National Archives at College Park, Maryland, Still Pictures Branch)

A USAF F-4 drops a B-83 bomb during the final flight test of the B-83 test programme in 1983. (US National Archives at College Park, Maryland, Still Pictures Branch)

A USAF F-4, armed with a low-altitude dispenser bomb, in flight in 1984. (US National Archives at College Park, Maryland, Still Pictures Branch)

A Tactical Air Reconnaissance Pod System on a USAF F-4. (US National Archives at College Park, Maryland, Still Pictures Branch)

A USAF F-4D, carrying a High-Speed Anti-Radiation Missile, in flight in 1983. (US National Archives at College Park, Maryland, Still Pictures Branch)

A practice AGM 45A Shrike missile on a USAF F-4. (US National Archives at College Park, Maryland, Still Pictures Branch)

USAFE 52nd TFW pilots perform pre-flight checks on Mark 20 Rockeye cluster bombs on their F-4 Phantom II in 1981. (US National Archives at College Park, Maryland, Still Pictures Branch)

A USAF F-4E, loaded up with an Air Combat Maneuvering Instrumentation pod
and reflective laser device, heads out on a mission from Osan airbase, South Korea,
in 1982. (US National Archives at College Park, Maryland, Still Pictures Branch)

USAF Load Crewchief Staff Sergeant Robert Ward, 107th Fighter Interceptor Group, mounts an SUU-23/A 20mm gun pod on an F-4 during Exercise William Tell '88. (US National Archives at College Park, Maryland, Still Pictures Branch)

An Air Force Reserve F-4D, carrying a gun pod, returns to Nellis AFB, Nevada, during Exercise Gunsmoke '85. (US National Archives at College Park, Maryland, Still Pictures Branch)

USAF 51st TFW ground crews service an F-4E during Exercise Pitch Black '84. (US National Archives at College Park, Maryland, Still Pictures Branch)

A 159th Tactical Fighter Group, Louisiana Air National Guard, ground-crew member services an F-4 Phantom II's radar. (US National Archives at College Park, Maryland, Still Pictures Branch)

A Hill AFB-based F-16 Fighting Falcon and F-4D Phantom II in flight in 1980. (US National Archives at College Park, Maryland, Still Pictures Branch)

Two USAF F-4Es take to the skies during exercise Team Spirit '82, while 3380th Security Police Squadron troops maintain a defence from an attacking force at Kunsan airbase, South Korea. (US National Archives at College Park, Maryland, Still Pictures Branch)

A USAF 3rd TFS F-4E in flight during exercise Cope North '81–4. (US National Archives at College Park, Maryland, Still Pictures Branch)

A formation of three USAF F-4E Phantom IIs and two USAF F-15 Eagles in flight during exercise Cope North '81–4. (US National Archives at College Park, Maryland, Still Pictures Branch)

Mark 82 bombs are prepped for loading onto a USAF F-4E at Osan airbase, South Korea, during exercise Team Spirit '84 – a joint US/South Korean exercise. (US National Archives at College Park, Maryland, Still Pictures Branch)

A USAF 3rd TFS / 3rd TFW F-4E heads out to perform a mission during Exercise Opportune Journey 4 in 1982. (US National Archives at College Park, Maryland, Still Pictures Branch)

A USAF 3rd TFS F-4E Phantom II, with drag chute deployed, performs a landing during Exercise Coral Sea '85. (US National Archives at College Park, Maryland, Still Pictures Branch)

A USAF F-4 Phantom II prepares to land at Clark Air Base, Luzon, Philippines, in 1985. (US National Archives at College Park, Maryland, Still Pictures Branch)

A USAF F-4 Phantom II undergoes aerial refuelling with a KC-10 Extender above Hawaii during Exercise Cope Canine '85. (US National Archives at College Park, Maryland, Still Pictures Branch)

A 199th TFS, Hawaii Air National Guard, F-4C, sporting four North Vietnamese Air Force MiG kill markings, in flight in 1979. (US National Archives at College Park, Maryland, Still Pictures Branch)

A USAF F-4 heads out on a mission from Osan airbase, South Korea, during Exercise Team Spirit '86. (US National Archives at College Park, Maryland, Still Pictures Branch)

Three USAF 497th TFS F-4Es peel off to the left during Exercise Team Spirit '86. (US National Archives at College Park, Maryland, Still Pictures Branch)

A US Navy F-4 Phantom II lands on the USS *Independence*, operating off the Virginia Capes, in 1989. (US National Archives at College Park, Maryland, Still Pictures Branch)

A VF-102 F-4J in flight. (US National Archives at College Park, Maryland, Still Pictures Branch)

Two Phantom IIs about to be catapult-launched off the USS *Midway* in 1984. (US National Archives at College Park, Maryland, Still Pictures Branch)

An F-4 is launched off USS *Midway* in 1984. (US National Archives at College Park, Maryland, Still Pictures Branch)

A VF-24 F-14A Tomcat and two VF-301 F-4Ss fly in formation above Yuma, Arizona, in 1986. (US National Archives at College Park, Maryland, Still Pictures Branch)

A USMC VMFA-321 F-4S undergoes maintenance at NAS Miramar, California, in 1987. (US National Archives at College Park, Maryland, Still Pictures Branch)

Side view of a USMC VMFA-321 F-4S in 1987. (US National Archives at College Park, Maryland, Still Pictures Branch)

Side view of another USMC VMFA-321 F-4S in 1987. (US National Archives at College Park, Maryland, Still Pictures Branch)

A USAF F-4G Wild Weasel in flight during Exercise Gallant Eagle '82. (US National Archives at College Park, Maryland, Still Pictures Branch)

A pair of USAF F-4Gs undergo aerial refuelling from a KC-135 Stratotanker during Exercise Gallant Eagle '82. (US National Archives at College Park, Maryland, Still Pictures Branch)

Two USAF F-4Gs in flight. (US National Archives at College Park, Maryland, Still Pictures Branch)

A pair of George AFB 563rd TFS F-4Gs perform a mission during Exercise Reforger '81. (US National Archives at College Park, Maryland, Still Pictures Branch)

A USAF 563rd TFS F-4G is prepped for another mission at RAF Wildenrath, West Germany, during Exercise Reforger '81. (US National Archives at College Park, Maryland, Still Pictures Branch)

A USAF F-4G in flight above South Korea during Exercise Team Spirit '82. (US National Archives at College Park, Maryland, Still Pictures Branch)

A USAF F-4G drops bombs on a practice target during Exercise Team Spirit '82. (US National Archives at College Park, Maryland, Still Pictures Branch)

US Air Force in Europe – F-4G, A-10A and RF-4C – perform a flight mission above West Germany in 1987. (USAF photo by SSgt. David Nolan)

F-4CCV Fly-By-Wire Technology Demonstrator

The F-4CCV flight research aircraft served the USAF in a variety of capacities. The aircraft began its life as an F-4B, but was later modified to serve as the prototype YRF-4C. In the years to follow, it was further modified to serve as the prototype YF-4E and it was then utilised by the USAF to serve as a fly-by-wire (FBW) survivable flight control systems (SFCS) technology demonstrator. The aircraft was later designated the Precision Aircraft Control Technology (PACT) demonstrator. It performed its maiden flight, with the assistance of back-up mechanical control systems, in the skies above Lambert St Louis International on 29 April 1972. However, on 22 January 1973, the PACT demonstrator flew for over an hour in the skies above Edwards AFB, California, relying solely on a newly installed FBW control system. For enhanced Control Configured Vehicle (CCV) studies, the aircraft was outfitted with special test canard foreplanes and flaperons. With these modifications, the aircraft successfully performed thirty research flights over Edwards AFB. At the conclusion of the flight research programme, the F-4CCV was retired and transferred to the National Museum of the USAF on 5 December 1978.

The USAF F-4 Control Configured Vehicle, a fly-by-wire technology demonstrator aircraft, in flight above Edwards AFB, California, on 12 August 1974. (US National Archives at College Park, Maryland, Still Pictures Branch)

The USAF F-4 Control Configured Vehicle banks to the right during a test flight above Edwards AFB, California, on 12 August 1974. (US National Archives at College Park, Maryland, Still Pictures Branch)

EXPORT PHANTOM IIs

The McDonnell Douglas F-4 Phantom II was one of the most widely exported US-made aircraft in aviation history, equipping the air forces of numerous countries around the world. Phantom II operators included Australia, Egypt, Germany, United Kingdom, Greece, Iran, Israel, Japan, Spain, South Korea and Turkey.

Australian Phantom IIs

Between 1970 and 1973, the Royal Australian Air Force (RAAF) acquired twenty-four USAF F-4Es. The aircraft were loaned to the RAAF and were slated to fill its need for an advanced fighter-bomber until it acquired its General Dynamics F-111Cs. The RAAF F-4Es were assigned to No. 1 and No. 6 Squadrons based at RAAF Amberley.

Egyptian Phantom IIs

Since 1979, the Egyptian Air Force has operated a fleet of forty-five ex-USAF F-4Es. The aircraft are equipped with Sparrow, Sidewinder and Maverick missiles.

An Egyptian Air Force 222nd Fighter Regiment F-4E forms up with a USAF 347th TFW F-4E during Exercise Proud Phantom in 1980. (US National Archives at College Park, Maryland, Still Pictures Branch)

German Phantom IIs

In January 1971, the German Air Force (Luftwaffe) acquired eighty-eight RF-4Es from the United States. These photoreconnaissance aircraft were modified in 1982 to perform ground-strike missions, but the ground-strike capable RF-4Es were mothballed in 1994. In 1973, the Luftwaffe acquired F-4Fs from the US as part of the Peace Rhine programme, and more advanced versions of the F-4F were acquired by Germany during the mid-1980s. The USAF 49th Tactical Fighter Wing, based at Holloman AFB, conducted training operations for Luftwaffe pilots, utilising twenty-four Luftwaffe F-4F Phantom IIs from the 1980s to December 2004. In 1983, Germany commenced the Improved Combat Efficiency (ICE) programme, which resulted in the addition of 110 advanced F-4Fs to Luftwaffe operational fighter squadrons in 1992. Most of the Luftwaffe's Phantom IIs were assigned to Jagdgeschwader 71 at Wittmund, located in northern Germany, and WTD61 at Manching. All Luftwaffe F-4Fs were mothballed on 29 June 2013. Over their service lives with the Luftwaffe, the F-4Fs logged a total of 279,000 flight hours.[13, 14]

Two German Luftwaffe Jagdgeschwader 74 (JG 74) F-4F Phantom IIs in flight above Germany in January 1998. (USAF photo by Technical Sergeant Brad Fallin)

A West German F-4F takes off from Ahlhorn airbase, West Germany, in 1984. (US National Archives at College Park, Maryland, Still Pictures Branch)

A West German F-4F lands at Wittmundhaven airbase, West Germany in 1989. Note German Luftwaffe Second World War fighter ace Erich Hartmann's F-86 Sabre jet fighter on display in the foreground. (US National Archives at College Park, Maryland, Still Pictures Branch)

A West German RF-4 Phantom II flying in formation with two USAF F-15 Eagles above West Germany during Exercise Reforger-Crested Cap II in 1982. (US National Archives at College Park, Maryland, Still Pictures Branch)

Hellenic (Greek) Phantom IIs

In 1974, the Hellenic Air Force acquired a small fleet of F-4Es from the US, and at the beginning of the 1990s, it obtained additional German RF-4Es and US Air National Guard F-4Es. By 2013, there were thirty-four operational advanced F-4E Peace Icarus 2000 Phantom IIs in service with the 338 and 339 Squadrons of the Hellenic Air Force, as well as twelve operational RF-4E Phantom IIs in service with the Hellenic AF 348 Squadron. All Hellenic AF RF-4Es were retired on 5 May 2017.

Iranian Phantom IIs

During the late 1960s and 1970s, Iran acquired 225 F-4D, F-4E and RF-4E Phantoms that were pressed into service with the Imperial Iranian Air Force. These Phantom IIs proved to be effective, serving with the Islamic Republic of Iran Air Force during the 1980s Iran–Iraq War. An

example of this combat effectiveness was demonstrated during Operation Scorch Sword, when two F-4s performed a successful strike mission on the Iraqi Osirak nuclear reactor close to Baghdad on 30 September 1980. In another instance of combat effectiveness, eight IRIAF F-4s raided the Iraqi H-3 air-base network in western Iraq. Consequently, numerous Iraqi Air Force aircraft were demolished on the ground with no IRIAF F-4 casualties being experienced. IRIAF Phantom IIs saw action again in 2014, when they successfully performed strike missions against ISIS forces in Iraq's Diyala province, located in the eastern portion of the country.

Israeli Phantom IIs

Perhaps the largest foreign Phantom II customer was Israel. The Israeli Air Force (IAF) first acquired former USAF F-4Es and RF-4Es as part of the Peace Echo I programme in 1969. In the years to follow, the Peace Echo II–V and Nickel Grass initiatives resulted in Israel's accumulation of more former USAF Phantom IIs. During the 1980s, Israel modernised its F-4 fleet through the initiation of the Kurnass (Sledgehammer) 2000 initiative. IAF Phantom IIs often carried Israeli-made Rafael Shafrir and Python heat-seeking missiles. The Shafrir established a 60 per cent kill rate throughout the various Arab-Israeli conflicts in which Israeli Phantom IIs have participated in.[15] All IAF Phantom IIs were mothballed in 2004.

Japanese Phantom IIs

Since 1968, the Japan Air Self-Defense Force (JASDF) has acquired 140 F-4EJ Phantom IIs. Ninety-six F-4EJs were upgraded to F-4EJ Kais, while fifteen underwent conversion to RF-4EJ reconnaissance variants. JASDF Phantom IIs often carried Japanese-made Mitsubishi AAM-2 heat-seeking missiles. As of 2007, there were approximately ninety Phantom IIs in JASDF operational service. In 2011, Japan agreed to replace its F-4s with the stealthy F-35 Lightning II. Today, only three Japanese Phantom II squadrons remain operational, all of which maintain operations from Hyakuri airbase.

An Islamic Republic of Iran Air Force F-4E undergoes aerial refuelling during a mission of the Iran–Iraq War in 1982. (IRIAF)

Two Japanese Air Self-Defense Force F-4EJ Phantom IIs take to the skies above Misawa airbase, Japan, in September 2002. (US Navy photo by John Collins)

Two Japanese Air Self-Defense Force (JASDF) F-4EJ Phantom IIs fly in formation with a USAF F-15 Eagle during the JASDF/USAF Exercise Cope North 83–4 in 1983. (US National Archives at College Park, Maryland, Still Pictures Branch)

Another view of the previously depicted aircraft. (US National Archives at College Park, Maryland, Still Pictures Branch)

A flight of a Japanese F-1, a Japanese F-15, a USAF F-15, a Japanese RF-4 and a Japanese F-4EJ in 1984. (US National Archives at College Park, Maryland, Still Pictures Branch)

South Korean Phantom IIs

In 1968, the Republic of Korea (South Korean) Air Force acquired several ex-USAF F-4D Phantom IIs through the Peace Spectator programme. South Korea continued purchasing F-4Ds from the US until 1988. It later obtained ex-USAF F-4Es from the US as part of the Peace Pheasant II programme.

Spanish Phantom IIs

In 1971, the Spanish Air Force obtained several former USAF F-4C Phantom IIs, known as C.12s, as part of the Peace Alfa programme. The C.12s were mothballed in 1989, but, simultaneously, Spain also acquired several former USAF RF-4Cs, which became known as CR.12s. All CR.12s were retired in 2002.

A South Korean F-4E, loaded up with an AGM-65A Maverick air-to-ground missile, in flight in 1979. (USAF photo by MSgt Philip J. Lewis)

Turkish Phantom IIs

In 1974, the Turkish Air Force (TAF) acquired forty F-4Es from the United States, and from 1977–78, it obtained an additional thirty-two F-4Es along with eight RF-4Es from the US as part of the Peace Diamond III programme. Then in 1987, it acquired yet another forty former USAF F-4Es, followed in 1991 by another batch of forty former US Air National Guard Phantom IIs. From 1992 to 1994, it acquired another thirty-two RF-4Es from Germany. Israel Aerospace Industries (IAI) undertook a modernisation effort on fifty-four TAF F-4Es in 1995, which became known as the F-4E 2020 Terminator. Operating jointly with Turkish F-16s, the Turkish Phantom IIs performed numerous successful ground-attack missions against Kurdish PKK targets in Northern Iraq. Turkey continues to use its F-4E 2020s to combat ISIS and the Kurdish PKK in Iraq, most recently in attack missions on 12 January and 12 March 2016.

British Phantom IIs

The United Kingdom purchased two Phantom II export variants from the United States that utilised the US Navy F-4J as their base design. The first variant, the F-4K (also known as the Phantom FG.1 (fighter/ground attack)), was intended for Royal Navy Fleet Air Arm use. The second variant, the F-4M (also known as the Phantom FGR.2 (fighter/ground attack/reconnaissance)), was intended for Royal Air Force use. Both of the British Phantom II variants featured new British Rolls-Royce Spey engines. Afterburning variants of these less problematic engines were ultimately incorporated in the designs of all British Phantom IIs. The Rolls-Royce Spey engines proved to be more powerful than their American counterpart engines utilised in the designs of US Navy and Marine Corps Phantom IIs. At first, British Phantom IIs carried American-made AIM-7E2 radar-guided missiles, but later carried British Aerospace Dynamics Sky Flash radar-guided missiles. The Sky Flash radar-guided missile was essentially the same as an American-made Sparrow missile, but featured a new guidance and control system that utilised a solid-state monopulse seeker. The Sky Flash missile proved to be an upgrade over the American-made Sparrow missile and better resisted electronic countermeasures.

On 30 April 1968, No. 700P Squadron became the first Royal Navy squadron to operate the Phantom II, primarily for trials duties, at Yeovilton. Following the naval Phantom II trials, the No. 767 Squadron was established for training purposes. Britain's tight defence budget meant that only one Royal Navy Phantom II squadron could be deployed for sea duty. As a result, No. 892 Squadron was detached to the USS *Saratoga*, on station in the Mediterranean Sea, in late 1969 and was assigned to both HMS *Ark Royal* and HMS *Eagle* in 1970. All Royal Navy Phantom IIs were decommissioned in 1978 and reassigned to RAF service. Within the Royal Navy, F-4Ks were replaced by Sea Harriers for fleet defence duty.

The first RAF Phantom IIs entered service with No. 6 Squadron, based at Coningsby, Lincolnshire, in May 1969. FGR.2s entered operational service as ground-attack and reconnaissance aircraft serving with the RAF in Germany, and were intended to serve as replacements for older Hawker Hunters. RAF 43 Squadron was equipped with FG.1s that provided air defence. By the mid 1970s, the majority of Phantom IIs serving as ground-attack fighter-bombers based in Germany were transferred back to England, where they replaced obsolescent English Electric Lightning fighters as interceptors. The RAF 111 Squadron was established in 1979 using FG.1s obtained from the disbanding of 892 NAS. FGR.2s were also operational with No. 54 Squadron at Coningsby. Ground-attack Phantom IIs in RAF service were eventually replaced by Jaguar strike aircraft.

The Phantom FGR.2 gained distinction in 1982 when three of these Phantom II variants, serving with RAF No. 29 Squadron, were deployed to Ascension Island during the Falklands War to serve as interceptors. The Phantom IIs performed so well during the war that fifteen modernised former USN F-4Js, designated F-4J(UK)s, were acquired by the RAF, with one FGR.2 squadron being permanently based at Port Stanley, Falkland Islands, to provide air defence. There were a total of fifteen RAF units flying a variety of Phantom II variants, most of which operated from bases in Germany. RAF Leuchars No. 6 Squadron became the first operational RAF Phantom II unit in July 1969. RAF No. 43 Squadron, stationed at Leuchars, happened to fly the Phantom FG.1 for twenty years from 1969 to 1989. During the late 1980s, air-defence Phantoms were supplanted by Panavia Tornado F3s. All British Phantom IIs were mothballed in October 1992.

The initial British Royal Navy F-4K prepares to land at McDonnell Aircraft Corporation, St. Louis, Missouri, on 28 June 1966. (US Navy)

A Royal Navy 892 Naval Air Squadron Phantom II FG.1 preparing for flight operations aboard HMS *Ark Royal* on 2 March 1972. (US Navy)

A Royal Navy Phantom II FG.1 undergoing carrier trials aboard HMS *Ark Royal*.
(US Navy)

An 892 Naval Air Squadron F-4K and VF-101 F-4J prepare for catapult launches aboard USS *Independence* in 1975. (US Navy)

An 892 NAS F-4K is launched from USS *Independence* in 1975. (US Navy)

An RAF 43 Squadron Phantom II FG.1 prepares to undergo aerial refuelling from a USAF KC-135 tanker in 1980. (USAF photo by Major Dennis A. Guyitt)

An RAF F-4M Phantom II FGR.2 undergoes flight-testing at NAS Patuxent River, Maryland, in 1970. Note the gun pod mounted under the fuselage for tests. (US Navy)

An RAF 92 Squadron F-4M lands at RAF Wildenrath, West Germany, during the 1980s. (Courtesy of Rob Schleiffert, via Wikimedia Commons)

An RAF 985/86 Interceptor Squadron Phantom II FGR.2. (Courtesy of clipperarctic, via Wikimedia Commons)

An RAF No. 74 Squadron F-4J (UK) or Phantom F.3 in flight in 1984. (US Navy)

A US Navy VF-32 F-14A Tomcat in flight with an RAF 19(F) Squadron Phantom FGR.2 in December 1990. (US National Archives at College Park, Maryland, Still Pictures Branch)

10

PHANTOM IIs IN OPERATION DESERT STORM

On 15 August 1990, a fleet of twenty-four USAF F-4G Wild Weasel Vs and six USAF RF-4Cs arrived at Shaikh Isa airbase, Bahrain, in support of Operation Desert Shield. The aircraft were intended to provide coalition forces with anti-SAM sites and long-range photoreconnaissance assets. The F-4Gs belonged to the 35th TFW, based at George AFB, California, and the 52nd TFW, based at Spangdahlem, Germany.

Following the commencement of Operation Desert Storm on 17 January 1991, the USAF F-4Gs and RF-4Cs performed missions on a daily basis, with the F-4Gs destroying numerous Iraqi SAM sites and

the core of the Iraqi air-defence network. During Desert Storm, the Phantom II Wild Weasels performed 3,942 combat missions and launched 1,000 air-to-ground missiles, destroying some 200 Iraqi SAM sites in the process.[16] During the initial days of Desert Storm, F-4Gs targeted the Al-Taqaddum airbase, located close to Fallujah. This Iraqi airbase had been known to operate advanced MiG-29 Fulcrum jet fighters and was also equipped with SA-2 and SA-3 missiles, making it a high-risk target. To perform their high-priority missions, the F-4Gs were equipped with three external fuel tanks, an AN/ALQ-119 electronic counter-measures

USAF 35th TFW F-4G Wild Weasels at an airbase in Saudi Arabia during Operation Desert Storm. (US National Archives at College Park, Maryland, Still Pictures Branch)

A pair of USAF 35th TFW F-4Gs in flight above the Saudi Arabian desert in 1991. (USAF)

(ECM) jamming pod, two AIM-7 Sparrow missiles for aerial combat and self-defence from aerial interception, and two AGM-88 High-Speed Anti-Radiation Missiles (HARMs). To complete their missions, the Wild Weasels had to undergo aerial refuelling twice. The US Navy also performed Wild Weasel missions during Desert Storm, but using their F/A-18 Hornets. By the conclusion of the First Persian Gulf War, the Phantom II Wild Weasels had eliminated 74 per cent of all Iraqi SAM site radars disposed of during the conflict.[17]

During Operation Desert Storm, only one F-4G became a casualty when the aircraft's fuel tanks were struck by Iraqi ground fire, causing the Phantom II to run out of fuel close to a coalition airbase. The deployment of Phantom IIs to Operation Desert Storm had proven to be a huge success, significantly contributing to coalition victory.

11

THE PHANTOM II's TWILIGHT AND RETIREMENT

The US Navy's F-4Ns were retired in 1984 and its F-4Ss in 1987. Another milestone in the Phantom II's illustrious history was achieved on 25 March 1986, when a VF-151 Vigilantes F-4S performed the final regular service US Navy Phantom II launch from a carrier (USS *Midway*); on 18 October 1986, the final Phantom II carrier landing was performed by a Naval Reserve VF-202 Superheats F-4S aboard *America*. The US Navy continued to operate QF-4 target drones until 2004. F-4 Phantom IIs were replaced by Grumman F-14 Tomcats and McDonnell Douglas F/A-18 Hornets in the air superiority role for the US Navy.

During the early 1980s, regular and reserve USMC fighter-attack squadrons converted from Phantom IIs to F/A-18 Hornets. With the retirement of VMFA-112's F-4Ss on 18 January 1992, there were no more Phantom IIs in service with the USMC.

The Grumman F-14 Tomcat eventually replaced the McDonnell Douglas F-4 Phantom II for the US Navy. (US National Archives at College Park, Maryland, Still Pictures Branch)

The McDonnell Douglas F/A-18 Hornet eventually replaced the McDonnell Douglas F-4 Phantom II for the USMC. (US National Archives at College Park, Maryland, Still Pictures Branch)

The sun sets on a ground-crew member performing maintenance on a USAF 37th TFW F-4 Phantom II in 1990. (US National Archives at College Park, Maryland, Still Pictures Branch)

On 26 March 1996, the last of the USAF Phantom IIs, 561st Fighter Squadron F-4G Wild Weasel Vs, were decommissioned. The final F-4G Wild Weasel mission was performed by the Idaho Air National Guard's 190th Fighter Squadron in April 1996. QF-4 target drones were in service with the USAF until 2015, when they were replaced by QF-16s. By December 2013, a total of 250 unmanned USAF QF-4s had been shot down. Phantom IIs were replaced by McDonnell Douglas F-15

A USAF QF-4 in flight in 2008. (USAF)

Eagles and General Dynamics F-16 Fighting Falcons in the air superiority and Wild Weasel roles for the USAF.

Throughout its illustrious career, the McDonnell Douglas F-4 Phantom II did it all. It began its career serving as a highly productive performance record setter, and it provided critical fleet defence coverage for the US Navy during the Cuban Missile Crisis, when the world came as close to a nuclear war as it had ever come. Later, during the Vietnam

The McDonnell Douglas F-15 Eagle eventually replaced the McDonnell Douglas F-4 Phantom II for the USAF. (US National Archives at College Park, Maryland, Still Pictures Branch)

War and the Arab-Israeli conflicts of the 1970s and 1980s, it served as the ultimate advanced fighter-bomber, racking up impressive victory tallies over Soviet-made MiGs in the air and taking out a multitude of high-priority, high-risk targets on the ground. During the Cold War, the Phantom II served as a prime Soviet long-range bomber interceptor and weapons-test platform. During the Falklands War, Phantom IIs provided crucial air-defence support for British forces on Ascension Island. Finally, the Phantom II flourished at the end of its career, serving as a Wild Weasel for the USAF during Operation Desert Storm, in which it decimated a large portion of Iraq's elaborate air-defence network. These accomplishments truly make the McDonnell Douglas F-4 Phantom II worthy of the title 'Air Superiority Legend'.

APPENDIX
SURVIVING PHANTOM IIs AND THEIR DISPOSITIONS

The following is a list of F-4 survivors and exhibits located throughout the world:

1. F-4C Phantom II, SCAT XXVII, in which Colonel Robin Olds and Lieutenant Stephen Croker downed two North Vietnamese MiG-17s, exhibited at the USAF Museum at Wright-Patterson AFB near Dayton, Ohio.

2. F-4E Phantom II, hanging from the ceiling in Virginia Air & Space Center, Hampton, Virginia.

3. USAF Museum F-4C Phantom II, 63-7699, exhibited at Midland Air Museum, Coventry.

4. US Navy F4H-1 Phantom II, Bu No. 145310, exhibited at French Valley Airport, Marietta, California.

5. F-4C Phantom II exhibited at the Museum of Flight, Boeing Field, Seattle. The aircraft downed three MiG-21s during the Vietnam War.

6. F-4 Phantom II, sporting six North Vietnamese MiG kills, exhibited at the United States Air Force Academy.

7. F-4 Phantom II exhibited at Luke AFB.

8. 12th TFW/111th TFS (Texas Air National Guard) F-4C Phantom II, 62-0712, exhibited at Camp Mabry, Austin, Texas.

9. Mothballed reserve F-4s in storage at Davis-Monthan AFB, Arizona.

10. F-4D flown by the Collings Foundation.

The Collings Foundation F-4D at Selfridge Air National Guard Base, Michigan, in 2005. (Jacobst at English Wikipedia, via Wikimedia Commons)

NOTES

1   Thornborough, Anthony M., and Davies, Peter E., *The Phantom Story* (London: Arms and Armour Press, 1994), p. 15.

2   RG 342-B, Vol. 1. Records of Air Force Commands, Activities, and Orgs, Aircraft (B+W), F4, JF-4C, F-4C, Prints: USAF Activities, Facilities and Personnel, Domestic and Foreign 1954–1980, 01–051, Box 51, Folder 342-B-01-051-1 'F-4', US Department of Defense News Release No. 158–63 Oxford 75131, 5 February 1963, 'Statement by Lt. Gen. Gabriel P. Disosway, USAF, DCS/ Programs and Requirements' (US National Archives at College Park, MD, Still Pictures Branch), pp. 1–3.

3   Richardson, Doug, and Spick, Mike (1984), *Modern Fighting Aircraft, Volume 4: F-4* (Published in United States by Arco Publishing Inc., New York, NY, and in United Kingdom by Salamander Books Ltd, London), p. 12.

4   Chambers, Joseph R. (2000) NASA SP-2000-4519, *Partners in Freedom: Contributions of the Langley Research Center to U.S. Military Aircraft of the 1990's*, Monographs in Aerospace History Number 19, The NASA History Series, National Aeronautics and Space Administration, Washington, DC, p. 175.

5   Grossnick, Roy, and Armstrong, William J., *United States Naval Aviation, 1910–1995* (Annapolis, Maryland: Naval Historical Center, 1997).

6   Kirk, R., & Lihani, R. (producers) (8 February 2009), Dogfights 'Supersonic' (Transcript, Television series episode), in Dogfights. Houston, Texas: The History Channel.

7   'Phil Handley – FU Hero', Fighterpilotuniversity.com, 6 September 2008: www.fighterpilotuniversity.com/history/ fu-heroes/phil-handley-fu-hero.

8   Correll, John T., 'The Vietnam War Almanac', in *Air Force Magazine*, September 2004. See also USAF Operations Report, 30 November 1973. Retrieved: 19 November 2007.

9   Ibid.

10 Ibid.

11 Richardson, Doug, and Spick, Mike (1984), *Modern Fighting Aircraft, Volume 4: F-4* (Published in United States by Arco Publishing Inc., New York, NY, and in United Kingdom by Salamander Books Ltd, London), p. 56.

12 Ibid, pp. 56–7.

13 Hoyle, Craig, 'German air force to bid "Pharewell" to last F-4Fs', Flightglobal.com, 26 June 2013: www.flightglobal.com/news/ articles/picture-german-air-force-to-bid-pharewell-to-last-f- 4ts-387655.

14 'Auf Wiedersehn, Phantom!', *Aviation Week* blog, 1 July 2013.

15 Richardson, Doug, and Spick, Mike (1984), *Modern Fighting Aircraft, Volume 4: F-4* (Published in United States by Arco Publishing Inc., New York, NY., and in United Kingdom by Salamander Books Ltd, London), p. 37.

16 Dorr, Robert F., 'Gulf War 20th: Desert Storm Was the First and Last War for the F-4G Advanced Wild Weasel', Defense Media Network, 16 January 2011: www.defensemedianetwork.com/ stories/gulf-war-20th-desert-storm-was-the-first-and-last-war-for- the-f-4g-advanced-wild-weasel.

17 Ibid.